The Basic Chemistry of Aromatherapeutic Essential Oils

by E. Joy Bowles

Published by E. Joy Bowles, Sydney, Australia.

©1991, 1993, 1997, 2000 by E. Joy Bowles

First published 1991
Second revision 1993
Third revision 1997
Second Edition 2000

Printed by Pirie Printers.
Cover design and layout advice by Animae.

ISBN 0 6646 29413 X

Acknowledgements

At the time of first writing this book, I had no idea that it could become a standard basic text on the chemistry of essential oils for aromatherapists. I acknowledge the following people who were instrumental in encouraging me to write the book and design the accompanying course, as without them, nothing would have begun: Dr Lyall Williams and Pam Taylor for giving me the vision, John Fergeus, Paula Lipman, Peter Sayers and Gillian Kerr for helping with getting the first edition printed, and my students throughout the years for their input to the varying incarnations of the course.

As with all creative labours, this book has required much energy, grace and generosity from my friends, family and colleagues, and for everyone's gifts of soul I am deeply thankful.

Special acknowledgement must be given to Dr D. Pénoël and Pierre Franchomme, whose seminal work "L' aromathérapie exactement", set the foundations of essential oil chemistry for me.

It is my hope that this book will aid communication between aromatherapists and other health professionals, as each have valuable approaches to understanding the therapeutic principles of the essential oils.

E. Joy Bowles, Sydney, Australia.
June 2000

CONTENTS

THE BASIC CHEMISTRY OF AROMATHERAPEUTIC ESSENTIAL OILS

Why study the chemistry of essential oils?

Aromatherapy as a complementary therapy is growing at a rate unforeseen by most medical practitioners in mainstream medicine. The reasons for this are complex. People in today's highly pressured society are contracting new forms of disease, several of which appear to have no direct relation to an invasion by micro-organisms; and mainstream pharmacology has few answers to offer them. People suffering from myalgic encephalomyelitis (M.E.), multiple sclerosis, and AIDS, are often given very little hope of complete recovery. Many of them turn to alternative medicine and complementary therapies as a last straw of hope, and many of them find healing and complete restoration of their lives.

In Australia, alternative medicine is proceeding towards maturity as a serious option for health care. Moves are already underway to gain university accreditation for naturopaths and other practitioners. For aromatherapy to be included as a university subject, the more firmly grounded it is in a scientific paradigm, the better.

Aromatherapy already complies with certain basic scientific criteria such as the building up of empirical evidence and forming theories that explain the evidence. However, as the general public is better educated these days, it is no longer sufficient to say "Lavender oil will cure your headache because it has been used that way for the last few centuries." There is a much more critical and enquiring expectation from clients and doctors alike.

Consider the following scenario:

> *A woman visited you once for an aromatherapy massage, feeling depressed and stressed. You used Clary Sage oil on her, telling her that it would lift her spirits. She only had one session, and you didn't hear from her after that. Twelve months later, a doctor calls you from a hospital to tell you that this same woman has been admitted in premature labour, saying that maybe it was the essential*

oils she has been using that triggered the labour. The doctor is calling you to ask you about the properties of essential oils, and which ones are considered toxic. As the discussion continues, the doctor reveals that her patient has been using Sage oil repeatedly over the last six days in liberal quantities, mistakenly thinking that Sage and Clary Sage have the same anti-depressant properties.

With an understanding of chemistry, you can quickly check the constituents of Sage oil, and tell the doctor that Sage oil contains both the ketones thujone and camphor, and the oxide 1,8-cineole, all of which have neurotoxic effects at high enough doses[1], possibly triggering the contractions.

The purpose of this book is to give an introduction to the chemistry of the constituents of essential oils. It starts with a basic explanation of the scientific world view and a simplified explanation of the atomic and molecular structure of essential oils, followed by how plants make essential oils, the different types of molecules that are found in essential oils, and some pointers towards how the essential oils actually effect the human body therapeutically (or otherwise).

---

[1] R Tisserand & T Balacs, "Essential Oil Safety" Chapter 11, (1995) Churchill Livingstone

CHAPTER ONE

A SCIENTIFIC VIEW OF THE UNIVERSE

One of the tasks of being human is to come to an understanding of the universe we find ourselves in. The scientific perspective is to look at the components of the world, and observe how different components relate to each other, according to certain patterns which are called "laws". Looking at components is a fairly reductionistic way of thinking, but it serves the purpose of being able to explain physical phenomena, and gives us models to predict how new situations are going to occur.

From a simple dualistic point of view, everything we can perceive with our physical senses is either an arrangement of matter, or a direct manifestation of electromagnetic energy. Matter also has a certain amount of energy associated with it (such as bonding energy, thermal energy), and is affected by the energy of other matter, and by free electromagnetic energy, such as heat and light. Chemists are concerned with the way in which matter reacts with matter, and with the energy exchanges which occur during such reactions.

Matter, as we perceive it with our bodily senses, is what might be called the "stuff" of which our physical world is made. We can perceive different patterns in matter, such as the shape of crystals, the branching of fern leaves, the ripples in a pond, and we classify different types of matter into groups which share similar characteristics to help us make sense of our world.

On the scale of atoms and molecules, which is beyond the perception of our physical senses, each type of matter has a different structure. This atomic structure defines the characteristic properties of substances which we can perceive with our physical senses. Water, wood, gold, and air, are only different from each other because of their different atomic and molecular structures, though they are all made of matter.

Energy affects matter in different ways, depending on the sort of energy it is. In general, when energy interacts with matter, the matter is transformed, either physically, or chemically. Either way, the matter looks different after the energy has caused the structural change. For example, if you heat an ice-cube (heat is one form of electromagnetic energy), it will melt to become water and finally, steam. Ice, water and steam appear to be very different to our senses, but their atomic and molecular structures are still the same. This is an example of how energy changes the physical state of matter.

An example of how energy changes the chemical state of matter is to consider a wax candle burning. The energy that is supplied is in the form of the flame from a struck match[2] . The first change you notice is a physical change as the wax begins to melt. The chemical changes that are actually occurring, and producing the flame, are the combination of the wax with oxygen from the air to form two colourless gases, steam and carbon dioxide. There is also a little black soot (pure carbon) which is usually formed as well.

The important thing to notice is that although heat energy was responsible for both types of change - both the physical melting of the ice cube, and the conversion of wax into steam and carbon dioxide, only the physical changes were reversible. For purposes of this book, chemical changes are by and large permanent.

To understand how essential oils work as therapeutic agents in our bodies, we have to appreciate firstly that our bodies are a vast set of systems of chemical reactions. The addition of essential oils into the body allows for modification of those chemical reactions, resulting in toxicity or improved well-being[3].

The second step is to grasp the atomic and molecular structure of essential oil molecules, so as to be able to classify and predict their likely effects in the body. Understanding the body as a chemical system would require many years of study in pharmacology and biochemistry, and would give students a deeper understanding of how the essential oils affect the body. In this book I will highlight the effects of particular categories of essential oil molecules on specific body systems, supported with research examples where possible.

---

[2] You can trace the energy flow back through the friction of the striking, to the muscular energy of the arm, to the chemical energy of the food being converted into ATP, back through to the energy of the sun which the plants use to create the food we and other animals eat.

[3] The etheric effects of essential oils have been experienced by the author, but are beyond the scope of this book.

THE STRUCTURE OF ATOMS

As with all new subjects it is probably helpful to start with the most simple concepts and move towards the more complex. The pattern of using small units joined together to form larger units, which in turn can be joined together to form even larger structures is one which occurs on many different levels in our experience of life. When we come to look at matter, the "stuff" of which we are made, this is also the case.

The most useful concept to grasp when coming to understand the structure of matter is the concept of atoms. Atoms are made up of three types of subatomic particles of matter, which are positively charged protons, neutral neutrons and negatively charged electrons. Charge is a type of energy which is associated with these very small bits of matter, and gives them specific ways of relating with each other. As with magnets, where you have two opposite poles, a positively charged proton will attract a negatively charged electron, and repel another proton. Opposite charges attract, similar charges repel. The neutrons are not affected either way by the charge energy.

These charged particles are arranged within atoms in specific ways:
- The positive protons are in the centre or nucleus of the atom.
- The neutral neutrons are in the centre also, to prevent the positive charges from repelling each other.
- The negative electrons travel around the central nucleus in specific energy pathways (orbitals)
- Usually, the number of protons in an atom equals the number of electrons and the total charge on the atom, when added together, is zero.

Below are two ways of representing the nucleus and electron orbitals of atoms. The dots are the subatomic particles.

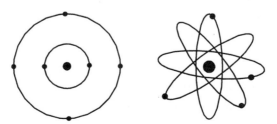

TYPES OF ATOMS FOUND IN ESSENTIAL OILS

Essential oils are manufactured by plants, and as with most organic plant substances, their starting materials are carbon dioxide and water. The component atoms of these two substances are hydrogen, carbon, and oxygen. A few essential oil constituents contain sulfur (e.g. diallyl sulfide, from garlic and the onion family), and some contain nitrogen (e.g. indole, from jasmine), but we will look only at hydrogen, carbon and oxygen here.

The structural difference between these types of atoms is all to do with numbers of subatomic particles (protons, neutrons and electrons) they contain. It is important to realise here that a proton from the centre of a gold atom is exactly the same as a proton from the centre of an oxygen atom. The only thing that makes the difference between a gold atom and an oxygen atom is the number of protons, neutrons and electrons they each have.

The first and smallest atom that exists is hydrogen. It has only one proton, and we would expect it to be accompanied by one electron. It is unique among atoms in that it doesn't need neutrons in its nucleus. The electron has energy associated with it which sends it spinning around the proton in the first orbital path, rather than being held in a tight embrace in the centre of the atom as you might expect due to the charge attraction between the two particles. It is somewhat like a horse being trained on a lunging rope - the horse (electron) goes round and round because it is restrained by the rope and the will and strength of the trainer (the charge attraction).

The next atom we concern ourselves with in this book is carbon. As it has six protons, it will have six electrons. There are also six neutrons in the center or nucleus of carbon atoms, which act as a kind of "bubble wrap" or insulating layer to prevent the six positively charged protons from repelling each other. As with hydrogen atoms, the electrons are all spinning in orbits around the nucleus, held by attraction to the protons, but impelled to whirl by their thermodynamic energy.

Oxygen atoms (our third type of atom), have eight positively charged protons and eight neutral neutrons in the nucleus, and also eight negatively charged electrons in spinning in orbitals around the nucleus.

But now we get to another stage in our understanding. The distribution of the positive and negative charges in the atom is quite specific, involving the grouping of the positive charges in a tight bundle in the centre of the atom, with the negative charges travelling in discrete energy paths around the positive charges. Negative charges in the path closest to the positive charges, will have the lowest energy, ones further away will have higher energies. If an atom receives sufficient free energy from outside (like heat or radiation), it can lose one or more of its negative charges. This would be as if the horse managed to have enough energy to break the lunge rope and be able to leap the fence and gallop away.

The most important function of these outer negative charges is that their movement gives us the sensation of solidity. A way of picturing it is a yo-yo being whirled on a string. Most of the time, in any one spot on the circumference of the path, there is empty space, but if you put your hand in the way of the whirling yo-yo, it soon wouldn't feel like empty space.

One characteristic of matter is that it will always tend to move towards a state of lowest energy, to be in the shape which requires least energy to maintain it, to be closest to the equilibrium position. Thus, when you have several negative charges, they will fill up the energy paths closest to the centre, or rather which have the lowest energy.

It has been discovered that the lowest energy path can contain only two negative charges. The second energy path can hold eight maximum, and the third also can hold eight. There are many more energy paths of higher energy, but for our purposes, we need only consider the lowest three.

BONDING

Although atoms have a balance of positive and negative charges, and are thus neutrally charged, most atoms do not exist by themselves in nature as single atoms. They bond together with other atoms to form molecules (the next level of structure).

The reason they do this is to achieve the stability which comes from completely filling their outermost orbital with electrons. As considered above, the first orbital of every atom can contain two electrons, and the second orbital can contain eight.

Examples of atoms which do exist by themselves in nature are helium and neon. Helium has 2 protons, 2 neutrons and 2 electrons, and as such is completely stable and no need for any other atom (a "self-actualised" atom!). Neon likewise has 10 protons, 10 neutrons and 10 electrons, which means that each neon atom has a full first orbital (2 electrons) and a full second orbital (8 electrons). Atoms which don't have full outer shells will bond with other atoms so as to achieve a sense of this between them.

Bonding between hydrogen, carbon and oxygen atoms is known as covalent bonding, which is sharing of electrons between the atoms[4]. In order for the atoms to share their electrons they actually have to get close enough to overlap their outer electron orbitals. This usually requires the presence of outside energy to get them moving fast enough to have the necessary energy to interpenetrate each other!

For our three atoms, the following table will help to explain how they bond together:

Atom	No. of electrons in outermost orbital	Bonds needed for molecular harmony
Hydrogen	1	1
Carbon	4	4
Oxygen	6	2

When two or more atoms are bonded together, they are called a molecule. Molecules act as if they are units in themselves, and have their own distinct properties. For example, hydrogen, an explosive gas, will bond with oxygen, the gas we need for life, to make water, which is a liquid. Water molecules do not act like either of their gaseous parent atoms.

Each overlap of electron orbitals (also known as a bond), enables the atoms involved to share a pair of electrons, one electron from each atom. Thus, for hydrogen to achieve bond harmony, it need only form one bond. Carbon atoms, on the other hand require four bonds to achieve molecular harmony. Oxygen atoms will form two bonds. Multiple bonds can be formed between two atoms, as long as the concentration of negative charges in one local area does not prove to be too great an energetic strain. The greatest number of bonds between two atoms is three.

---

[4] We will not consider ionic bonding as it is beyond the scope of this course, and may confuse people. Those wishing to explore further could get a high school chemistry text to start with.

When molecules are broken up into smaller molecules or atoms, the energy that was needed to force them to overlap in the first place is released as free energy again. The effect of bonding on the three dimensional structure of the molecule is very important, as the combined outermost energy paths of the atoms make a new molecular energy path. It is this molecular energy path which is characteristic of each individual type of molecule, and which gives it its properties.

DRAWING ATOMS AND MOLECULES

The language of chemistry is a curious mixture of letters and symbols, which together represent the relationship of atoms in molecules. This allows for a two dimensional representation of the three dimensional reality, and also provides a quick way of communicating about different molecules. Atoms are usually represented by capital letters, though some which were discovered by the alchemists are represented by the first two letters of their alchemical name e.g. Au is the symbol for gold from the name Aurum. The symbols of the atoms we are looking at are: hydrogen = H; carbon = C; oxygen = O.

As we have seen, there is a way of representing atoms using dots for subatomic particles and curved lines for electron orbitals, but for large molecules, this procedure would get very tedious. The simplest way to represent a molecule is by drawing sticks or bars between the symbols for the atoms, each stick representing one bond, the sharing of one electron from each atom.

Water molecules which contain one atom of oxygen and two atoms of hydrogen can be represented like this:

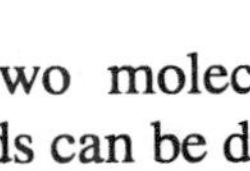

$$\overset{\displaystyle O}{\underset{H \quad\quad H}{\diagup \quad \diagdown}}$$

Carbon dioxide molecules (two molecules of oxygen, one atoms of carbon), with their double bonds can be drawn like this:

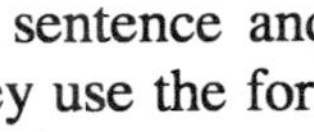

$$O=C=O$$

When chemists are writing a sentence and want to refer to the atomic components of a molecule, they use the formula of the molecule. This is an even shorter way of expressing types and numbers of atoms present.

For example, water is written H_2O, and carbon dioxide would be written CO_2. The small numbers indicate the number of atoms of each type present (e.g. H_2 means there are two atoms of hydrogen present).

For more complex molecules, there are a variety of styles of writing formulae. Let us take ethanol, the substance found in wine which inebriates one (referred to as "alcohol" by the general public!). It contains 2 carbon atoms, 6 hydrogen atoms and 1 oxygen atom. Its drawing would be:

$$
\begin{array}{ccccc}
 & H & & H & \\
 & | & & | & \\
H- & C & - & C & -O \\
 & | & & | & \backslash \\
 & H & & H & H
\end{array}
$$

According to convention, in written formulae of the most simple type, carbon is mentioned first, hydrogen second, and oxygen last. Thus you would have C_2H_6O. However, because this does not give us much information about the structure of the molecule, sometimes it is written CH_3-CH_2-OH. If you look at the stick structure, you can see how the carbon atoms form the skeleton of the molecule. If all you are interested in is the placement of the oxygen atom in the molecule, you can write it: C_2H_5OH.

In three dimensions, the molecules are usually not in straight lines, but the atoms will be joined to each other at angles which will ensure the maximum separation from each other. This is because the outer negative charges round each atom repel negative charges of other atoms. Thus, if a carbon atom has four other atoms (say, hydrogen atoms) joined to it, the three-dimensional shape of the molecule will be tetrahedral:

$$
\begin{array}{c}
H \\
| \\
H-C\cdots H \\
\backslash \\
H
\end{array}
$$

If one of the atoms joined to the carbon atom is doubly bonded to it, then the molecule will have a planar triangular shape. (Hydrogen cannot make double bonds as it only requires one additional electron to achieve molecular harmony.)

$$
\begin{array}{ccc}
H\backslash & & /H \\
 & C=C & \\
H/ & & \backslash H
\end{array}
$$

If the carbon atom bonds with another carbon atom, it is possible for them to share three bonds. The net result is a linearly shaped molecule:

$$H-C\equiv C-H$$

When dealing with molecules which have long carbon chains, such as most of the essential oil constituents, it becomes necessary to use an even more abbreviated form of representation. In the following example, each stick represents a single bond between two carbon atoms. The attachment of hydrogen atoms is presumed, and not shown, whereas the position of oxygen atoms is shown. The molecule is linalool, an essential oil constituent found in lavender oil and many other oils.

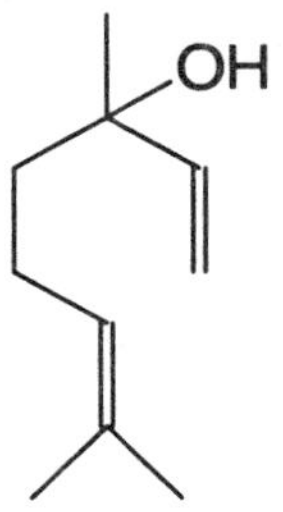

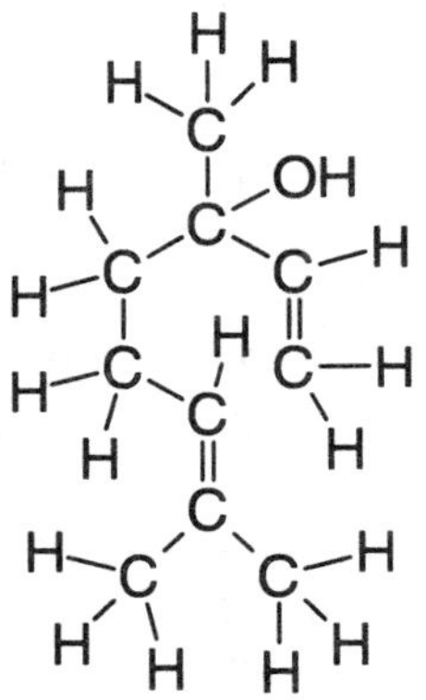

CHAPTER TWO

HOW PLANTS MAKE ESSENTIAL OILS

Having spoken of atoms, molecules and energy bonds, we are now in a position to get into the real world of how these molecules are actually made, and become the substances which we know as essential oils. It is all very well to know the structures and formulae of our constituents, but to gain an understanding of their place in our world, it is important to know how they are made.

Plants, like us, are made up of cells. These cells are themselves made up of atoms and molecules. The difference between molecules in living systems and those in non-living systems is that the molecules in living systems are continually in a state of flux. Most living molecules do not exist for very long in cells before they are radically changed by some exchange of energy, or some rearrangement of atoms.

A cell is a veritable laboratory, where millions of chemical reactions, or energy transfers are taking place every minute. Some of these reactions produce excess heat, which in the case of warm blooded animals is necessary for the other reactions to take place. Other reactions provide free energy in controlled amounts which allow particular molecules to combine their energy paths with other molecules to make different substances needed by the plant.

The main point is that cells contain the necessary energetic conditions for life. Cells are made up of sections, which have been likened to factory assembly lines, and it is in these, that the manufacture of particular chemical substances takes place.

In plants, there are particular cells which have the responsibility for collecting free energy for the plant to continue to live. Animals do not need to collect much free energy - they obtain most of their energy from food which they eat, having developed sophisticated systems for utilising the energy contained in food molecules.

The energy-collecting cells of plants are known as chloroplasts, because they contain a green substance known as chlorophyll (Chloro- = green;

phyll- = leaf, in Greek). This molecule has the characteristic that when it is hit by free energy in the form of light, two of its electrons get moved into a higher energy path than the remaining electrons of the molecule. When carbon dioxide (CO_2) and water (H_2O) molecules are held by molecular attraction close to the chlorophyll-containing sections in a leaf, a reaction can take place.

This reaction will involve the release of free energy by the chlorophyll molecule as the two electrons drop down into their usual energy pathway. The water molecule which receives this free energy, is broken up into hydrogen and oxygen atoms. Energy released from this reaction is sufficient to allow the energy paths of CO_2 and a specific enzyme to combine with the hydrogen and oxygen atoms from the water. A molecule three carbon atoms long is produced. When another such molecule is produced, the two can combine, to form a glucose molecule, which has six carbon atoms, twelve hydrogen atoms, and six oxygen atoms. This process is known as photosynthesis. (A L Lehninger in his book "Principles of Biochemistry" gives a more detailed account. Any final high school text book will give a simpler account.)

The free energy from the sun has thus been stored in the form of a complex molecule. The energy stored in the sugar molecules can then be released to provide energy for other chemical reactions that the plant needs to do. Because plants have neither mouths nor digestive systems (apart from some carnivorous plants which have modified structures to capture and digest insects), they rely on photosynthesis and uptake of nutrients and water through their roots for the manufacture of all the chemicals they need.

The "life-support" molecules that a plant needs to make are the same as for most living creatures. They need
- DNA and RNA (for their genetic code and protein manufacturing units)
- proteins (most enzymes are proteins or have a protein base)
- lipids and waxes (every cell has a cell membrane made of lipids, and the waxes coat the outside of leaves to prevent water loss)
- starches and vegetable oils (some plants choose to convert their excess glucose into these forms of energy storage molecules. Plants also need to make their support structures like stems, branches and roots. An example of a plant structural molecule is cellulose, a huge molecule made of tens of glucose molecules joined together, which give the crunch to celery sticks)

Plants have to solve another problem they have, which is that they can't run away from predators or other creatures that would threaten their lives. Some plants choose to make structures like thorns or spikes, but a large number of them choose to use chemical warfare. By producing chemicals which deter the predators, they can protect themselves for not as much risk as losing a branch or several leaves.

Some of the chemical warfare arsenal which plants create are the following

- terpenoid molecules (essential oil molecules are in this category)
- alkaloids (the "bitter", often toxic principles in herbs)
- flavanoids (found in fruits and vegetables - also play a part in plant hormones)
- saponins (make a froth when the plant is cut. The vegetable okra is a good example of a vegetable containing saponins, as you can taste the soapiness of the saponins.)

During the plant's life, all of these molecules play a role in the continuous dance of molecular interactions. They have different functions, some acting as chemical messengers from cell to cell, others going to make up the actual structure of the plant in the formation of new cells, still others being used in the defence of the plant against predators, or to attract insects for pollination purposes. They all share the characteristic of having been started off by a dose of free energy from the sun.

INTRODUCTION TO BOTANY

In this section of the book I have chosen to study members of the plant Family Labiatae (Lamiacae). Many of the more familiar essential oils come from this family, as do several cooking herbs. Examples are lavender, basil, hyssop, peppermint, rosemary and sage.

The study of botany is designed to provide the student with the ability to identify plants according to a pre-determined classification system. This system varies when new species of plants are discovered, and familiar ones are re-classified according to new criteria. For this reason it is important for aromatherapists to have an understanding of the botanical origins of their essential oils.

The classification system contains several categories each of which have an increasing specialisation. The categories are: Kingdom, Phylum, Class, Order, Family, Genus, Species. The most useful categories for aromatherapists are the last three, which provide information about the likelihood of plants to share similar growing habits, and possibly similar chemical constituents.

An example from the Family Labiatae is the *Mentha* genus. Several members of the *Mentha* genus contain the chemical constituent menthol, and they all share the characteristic of being small herbaceous plants. The different species have different shaped leaves and different coloured flowers, but they follow the same basic pattern. A comparison can be made between *Mentha piperita* and *Mentha arvensis*. The *piperita* species has long slender leaves, whereas the *arvensis* species has broader and hairier leaves.

What follows is a brief look at the various topics in botany, focussing on examples from the Labiatae family.

Taxonomy

In order for plants to be studied effectively, it is necessary to classify them into groups with general characteristics in common. This is usually done with respect to the following major features of plant morphology (shape)[5]:

- Leaf structure, number, and position on the stem
- Shape and positioning of the flowers, and number of their component parts (petals, stamens etc).
- Structure of fruit.

For the family Labiatae these characteristic features are as follows:

- Simple or pinnate leaf structure; usually two, opposite each other on the stem.
- Four or five petals in the flower, the bottom one being larger, giving the "lip" from which the family gets its name (labiate means "lipped"); usually in dense clusters near the joining of stem and leaf, commonly appearing to be in dense whorls, fifth stamen rarely present.
- Fruit composed of four nutlets, growing above, and enclosed by the calyx.[6]

---

[5] Simon JE et al. (1984) *Herbs: an index and bibliography,* Archer Books, Hamden, Connecticut

[6] For a more complete key, see "Thonner's analytical key to the families of flowering plants", by R Geesink, A J M Leeuwenberg, C E Ridsdale and J F Veldkamp (1981) Leiden University Press.

Ecological variables and their effects

Most of the family Labiatae are herbs which have developed in the temperate regions of the Mediterranean, or in similar climates around the world. Thus their preferred temperature band is between about 7 and 24 degrees Centigrade. Apart from this common preference, each genus has its preferred variables, which are quite specific. For example, *Ocimum basilicum* (Basil) is best grown in the sun, in well-drained soil. *Salvia sclarea* (Clary Sage) requires a dry calcareous soil, yielding very little essential oil if the soil is too rich. *Melissa officinalis* (Melissa or Lemon Balm) on the other hand requires deep soil, and is sensitive to excessive or inadequate water levels. *Origanum marjorana* (Oregano) also requires fertile, loamy soils, and is cold-sensitive. *Mentha pulegium* (Pennyroyal) is found in humid, fertile and partly shady regions.

The major effects that a variation from ideal conditions brings, are a reduction in the quantity and quality of essential oil produced. This is because the essential oils are not primary metabolites, and are produced when the plant is not under stress.[7]

Harvesting time and its effect on yield and quality of essential oil plants

In a study done by D. Basker and E. Putievsky in 1978, various members of the Labiatae family were tested to see when the optimum time for harvesting occurred. The conclusions which they came up with were:

- The volatile oil content of the leaves increases with time, and also with the size of the leaf.
- If the maximum leaf yield was measured, it was at the same time for most species, late summer, but the maximum oil yield and composition varied from species to species. Thus it is necessary to know at what time in the plant's cycle the plant matter has been harvested.

[7.]This is not a hard and fast rule, as very little is known about the factors which initiate the enzymatic pathways required for the manufacture of essential oils. In some plants, the essential oils may be produced only when the normal metabolites used in good conditions are not present. This is a fascinating area of study, as all the reactions that take place in plant cells monitor and feed-back on each other, and the system of the production of essential oils is subject to cellular conditions, and may not be as simplistic as I have suggested.

An example of this is *Salvia officinalis* (Sage), which contains different amounts of alpha-thujone depending on when it is harvested. It contains more after it has flowered, so it is harvested before flowering, as alpha-thujone is one of the more dangerous constituents of essential oils.

Hybridisation and cross-fertilisation

The family Labiatae in particular, are prone to naturally undergo hybridisation, and this means that defining species and sub-species can be a difficult job. Hybridisation is where two different species cross-pollinate, and the off-spring is usually sterile. In the family Labiatae, the off-spring of hybrids is often fertile, which makes for great difficulty in determining when a new species is formed.

For example, there are over 300 species of the genus *Thymus*, and several sub-species and varieties and ecotypes. A sub-species exists when a particular plant will produce others like itself, usually with some marked difference between it and the parent species. Varieties tend to apply to plants that differ from the parent species with some unimportant feature, such as colour of the blossoms. Ecotypes are plants which appear to exhibit characteristics like those of a sub-species, but when their seed is grown under different ecological conditions, they revert to their parent type.

Another type of classification which is used these days is chemotype. Chemotype refers to sub-species which have the same morphological characteristics, but which produce different quantities of chemical constituents in their essential oil. One example is the *Ocimum basilicum* species. The essential oil can be made up of either 70% methyl chavicol and 25% linalool, or it is made up of 50% linalool and 15% methyl chavicol[8]. (There are other constituents which are in different quantities too, but these are the most significant.)

When buying essential oils, it is important to know which chemotype you are buying, as these marked differences in the quantities of constituents will affect the medicinal properties of the oil. The label will probably read something like this *"Thymus vulgaris thymoliferum"*, or *"Thymus vulgaris C.T. thymol"* . Sometimes you are unlucky enough to get *"Thymus vulgaris C.T. 4"*, which can either refer to the marketer's own code, or

---

[8.] In fact the linalool/methyl chavicol ratio can be almost anything, but these two are quite common chemotypic examples.

perhaps to the standard as recorded in the British or European Pharmacopoeias.

It is often useful to know the geographical and ecological origin of the essential oils, as this information may give you a hint as to the chemotype, if this is not stated on the bottle. For example, Basil oil from the Comoros Islands is the methyl chavicol chemotype, and has a very pungent aniseed odour, whereas Australian Basil oil has a sweeter more herbaceous odour and is the linalool chemotype. In terms of therapeutic properties of the two different oils, the methyl chavicol type Basil would be good as an antispasmodic for the gut and stimulating to the trigeminal nerve (due to the methyl chavicol content), whereas the linalool chemotype would be more sedating and relaxing due to the higher percentage of linalool. I would not recommend using the linalool chemotype of Basil in a study or alertness blend.

CHAPTER THREE

TERPENOID MOLECULES

Most of the constituents of essential oils are terpenoid molecules[9]. Other constituents are short straight-chain molecules, and heterocyclic compounds such as indole. Some authors refer to all essential oil molecules by the loose term "terpenes". In this book the term "terpenes" refers to terpenoid hydrocarbons without oxygenated functional groups, which means that they contain only hydrogen and carbon atoms. The name "terpene" comes from turpentine, which in turn comes from the Old French "ter(e)binth", which means "resin". Initially, terpenoid molecules were thought of as having resinous properties and the characteristic smell of pine tree resin.

However, as other plants apart from turpentine trees were found to produce terpenoid molecules, most of which did not smell like pine, it became more useful to look at the actual structure of the compounds. Terpenoid molecules are made up from branched five carbon units, known as isoprene molecules.

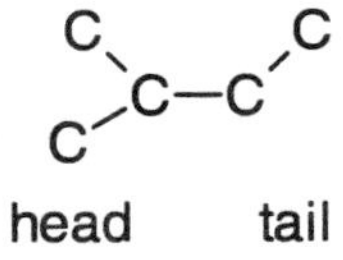

These units are joined "head to tail" to make up the variety of terpenoid molecules, and a few exceptions are joined "tail to tail"[10]:

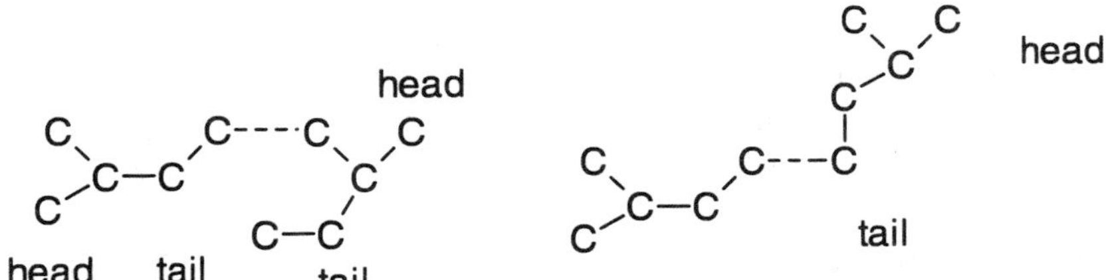

[9.]Phenyl propanoids are actually made by the shikimic acid pathway, but still look enough like terpenoids not to labour the distinction.

[10]These diagrams are taken from I L Finar (1975) "Organic Chemistry" vol.2, 5th Ed.

The two examples are monoterpenoid molecules, as they contain two isoprene units. Sesquiterpenoid molecules are made up of three isoprene units joined together, and diterpenoids have four isoprene units.

Physical characteristics of terpenoid molecules

Terpenoid molecules in essential oils share the characteristics of being volatile, flammable, less dense than water, and most of them are highly fragrant. There is a great range in thresholds of odour perception, with some substances like rose oxide (found in *Rosa damascena*) being detected at only 0.5 parts per billion (p.p.b.), and others like nerol, also found in *Rosa damascena*, which can only be detected at 300 p.p.b.[11].

If you were to ask the question about how terpenoid molecules behave when they come in contact with other substances, in particular the substances found in human bodies, such as water and various forms of fat and protein, you would have to look at the polarity and solubility of the molecules.

POLARITY

Going back to looking at molecules, most molecules usually have their bonding needs met, and the net charge over the whole molecule is zero because the numbers of protons and electrons are equal. However, when they are gathered together in a substance, some types of molecules exhibit an effect which is best explained by presuming that they are actually electrically charged. Water is one of these molecules, and surface tension and viscosity are related to this property of water. Machines which can measure the overall charge balance over a molecule of water detect that near the oxygen atom there is an area of slight negative charge, and near the two hydrogen atoms there are areas of slight positive charge.

Water molecules can thus be described as polar molecules because they have a negative and positive "pole", which behave like the North and South poles on a magnet - same charges repel, opposite charges attract.

When water molecules get together, the positive and negative ends link up forming loose liaisons (known as "hydrogen bonds" or "electrostatic liaisons", which are not covalent bonds as they are not sharing electrons). This forms a kind of lattice which apparently is what homeopathic remedies imprint energetically on (so a homeopath friend of mine

---

[11] Dodd GH (1988) "The molecular dimension in perfumery" Chapter 2 of *Perfumery: The psychology and biology of fragrance* Eds. Steve Van Toller & George H. Dodd, Chapman Hall, London p.33

suggested). It seems like the water molecules behave like members of an Old Girls' Reunion, holding hands, nattering together using the particular school girl jargon of their day, and generally offering the "cold shoulder" to anyone who cannot participate in the attractions and repulsions made possible by the polarity of the molecules (the old school tie!). Forming these electrostatic liaisons also allows the water to exist in a lower energy state, which is preferable.

Polar molecules

Molecules are polar when all these conditions are met:

- Their atoms are arranged asymmetrically, for example water:

    The unbonded electrons on the oxygen atom try to occupy the tetrahedral shape which has the lowest energy, i.e. is more stable.

- One or more of the atoms in the molecule attracts negative charges much more strongly than the other atoms do. Oxygen is the most electronegative of the atoms covered in this book. Electronegativity is the power of an atom to attract electrons to itself. (Think of it as a ratio between numbers of positive and negative charges - the 8 positive charges of the oxygen atom will have stronger attraction over the negative charges than the single positive charge in each hydrogen atom.)

- The highly electronegative atoms are arranged so that the net directions of attraction of electrons is not cancelled out, but added up. The direction of attraction of electrons is known as a dipole. (Draw arrows from the hydrogen atoms towards the oxygen atom on the water molecule above, and see that their net direction is an upwards one, rather than being equal and opposite).

Water is the only polar molecule which we will deal with in this book. Some essential oil molecules like indole (found in Jasmine) may be slightly polar due to the presence of moderately electronegative nitrogen atoms.

Non-polar molecules

Non-polar molecules are molecules in which the structure is symmetrical and there are no strongly electronegative atoms, or if there are, the dipoles cancel each other out.[12]

Terpenoid molecules are largely made up of carbon and hydrogen atoms, with the occasional oxygen atom present as part of a functional group. Because they are also symmetrical, and neither carbon or hydrogen are strongly electronegative atoms, there is no distribution of charge over the molecules. This means that terpenoid molecules (and other largely hydrocarbon molecules such as vegetable oils and fats) are largely non-polar.

Between all molecules in a substance, there exist other momentary attractions for the other molecules of that substance. This is what allows non-polar substances to become solids. These momentary attractions are caused by the continual movement of the negative charges in the molecular energy path (the composite energy path derived from the outermost energy paths of the atoms in the molecule). These attractions are very weak, and only act when the molecules are close together.

When non-polar molecules get together they mingle alongside one another, much like commuters on a rush-hour train. This is why essential oils dissolve in other non-polar substances like vegetable oils, mineral oils, full cream milk and in solvents used in cosmetics like isopropyl alcohol. Another term for these substances is that they are lipophilic, meaning they dissolve in fats, as opposed to hydrophilic substances which dissolve in water.

SOLUBILITY

The property of solubility is one of the most important for aromatherapists to understand, as it applies not only to the use of different solvents in blending, but also to the mechanism of epidermal penetration of the essential oils.

There are degrees of solubility, which are usually given either in a volume-to-volume ratio, (for example, 1:15 ethanol), or by the weight of substance that will dissolve in a given amount of solvent, for example, 2

[12.] Morrison RT & Boyd RN (1987) *Organic Chemistry,* Allyn & Bacon, USA.

g/L (grams per liter). Sometimes a description of the solution will be given, for example, "soluble to a turbid or opalescent state". Turbidity or milkiness is usually caused by the essential oil forming an emulsion of very small droplets, which is not a true solution.

The question "Why don't essential oils and water mix?" is one which is answered fairly simply, if you bear in mind that all molecules prefer to stay in the lowest energy state possible. As a general rule, the more complex a structure is, the more energy is required to hold it at that level of complexity.

So, when faced with a mixture of a polar substance such as water, and a non-polar substance such as an essential oil molecule, the following situations occur:

- Because the essential oil molecules are non-polar, the water molecules cannot form electrostatic liaisons with them. The presence of a non-polar molecule disrupts the low energy network of the water molecule electrostatic liaisons. So the water molecules try to form a "cage" around each essential oil molecule, so that it does not interfere with the normal process of hydrogen bonding in the rest of the body of water.

- The formation of such a "cage", however would take up too much energy, as it would have to be a complex structure, so the water molecules will rather repel the terpenoid molecules, refusing to let them enter the network at all.

- Because the terpenoid molecules are less dense than water, they float on the top of the water.

Most constituents of essential oils (and therefore, most essential oils) are not very soluble in water. The main reason is that they all have long non-polar carbon chains, which cannot dissolve in water, and thus the whole molecule is rejected.

Several essential oil molecules have an oxygen atom attached to a part of the molecule, which is known as the functional group. If the oxygen atom is part of a hydroxyl functional group (-OH group), this part of the molecule can form liaisons with water molecules. It is like half a water molecule itself, and the same conditions apply in terms of the electronegativity of the oxygen atom and subsequent arrangement of charges.

For a molecule such as a monoterpenol with 10 carbon atoms, a single - OH group is not sufficient to allow the molecule to dissolve in water. However, if a carbon-based molecule has several -OH groups, like glucose ($C_6H_{12}O_6$), it can readily dissolve in water, as the -OH groups act like a "passport" into the polar arena of the water. The diagram below shows menthol (a monoterpenol) and glucose.

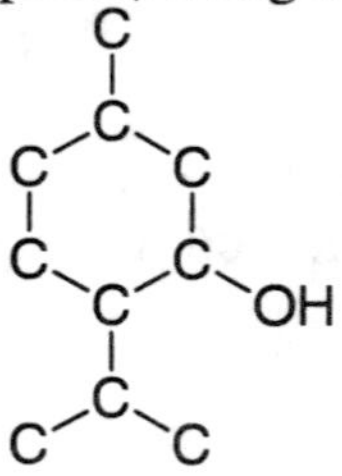

menthol 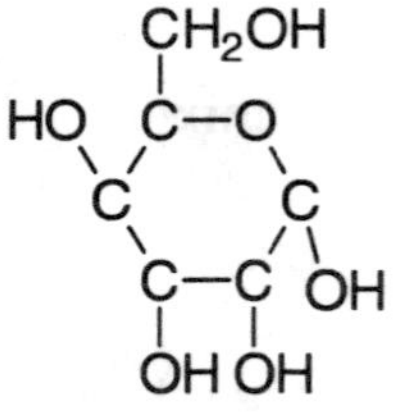   glucose

Essential oil constituents which have polar functional groups will dissolve a little bit in water, and thus the boundary between the oil and water layers will be less distinct. If there is sufficiently little essential oil, it is possible to achieve an emulsion by shaking vigorously, particularly if the oil is high in polar monoterpenoid constituents such as phenols, alcohols and acids. The larger the carbon chain, the less likely it is to be soluble in water.

The purpose of discussing polarity is to give an understanding of how essential oils might behave in a watery environment such as our bodies, in particular our blood and extracellular fluid. It would seem that although the oils might enter the epidermis, there would be little chance of penetrating much further due to the hydrophilic nature of the dermis.

However, if the oils penetrate a blood capillary near the epidermis, there are lipoproteins (e.g. LDLs (low density lipoproteins) and HDLs (high density lipoproteins)), which can carry the essential oil constituents on their cholesterol-based lipid chains. The hydrophilic protein portions of these molecules enable the whole molecule and the essential oil molecule to be solubilised in the blood.

Solubility of terpenoids in ethanol

Ethanol (commonly known as "alcohol" found in beverages) is a solvent often used to dissolve essential oils. It has a chain of two carbon atoms and a hydroxyl group. Ethanol dissolves in water because of the hydroxyl group (think of gin and tonic!), but also will dissolve some types of non-polar substances due to the carbon chain (see diagram below).

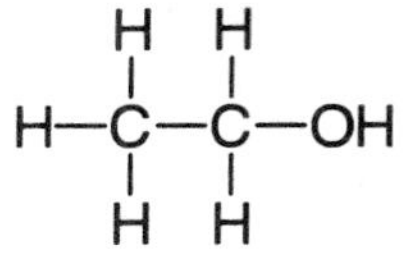

$$H-\underset{\underset{H}{|}}{\overset{\overset{H}{|}}{C}}-\underset{\underset{H}{|}}{\overset{\overset{H}{|}}{C}}-OH$$

ethanol

Most monoterpenoid constituents are soluble in ethanol. When you start having even longer carbon chains, such as in sesquiterpenoid or diterpenoid molecules, the mild polar inter-molecular forces of the ethanol molecules due to the -OH groups exclude the non-polar sesquiterpenoids.

CLASSIFYING TERPENOID MOLECULES BY THEIR STRUCTURE

The next section of this book will classify the different types of essential oil molecules, using their carbon chain structure, and the presence or absence of oxygen-containing functional groups as classification criteria. Each group of molecules also seems to exhibit some general therapeutic properties, which can be used as a rule of thumb when dealing with novel oils. The purpose of doing this is to provide the reader with an entry into the complex world of terpenoid chemistry, which will enable them to look at the chemical composition of any essential oil and begin to make sense of its potential as a therapeutic agent.

The following diagram over the page is like a road-map showing how the different types of essential oil molecules are derived from one another. The solid lines are definite chemical pathways, whereas the dashed lines involve complex rearrangements, and are only suggestions as to how the molecular structures are related. The fragments of the molecules shown are the parts of the molecules which are functional groups.

ESSENTIAL OIL CHEMISTRY ROAD MAP

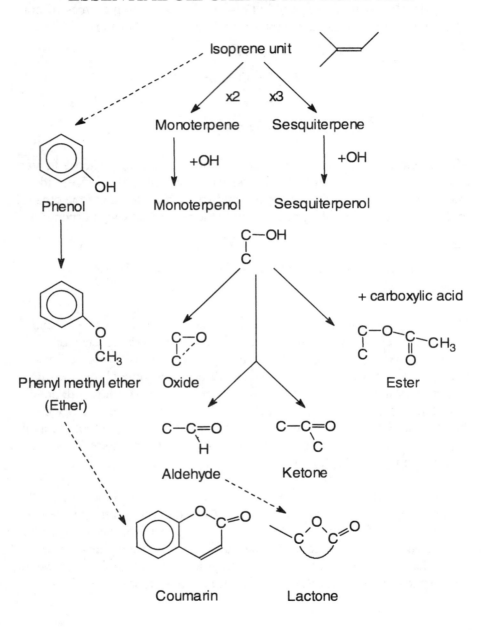

MONOTERPENES

Structure - Monoterpenes are hydrocarbons which have 10 carbon atoms (two isoprene units), with at least one double bond. The numbers of hydrogen atoms varies depending on the number of double bonds. They can have either open chain, monocyclic, or bicyclic structures.

Distribution of Monoterpenes in common essential oils

The values recorded in the tables below are average percentage values, rounded up to the nearest unit. The percentages come from the Boelens BACIS Essential Oil Database (1994).

Examples

alpha-myrcene $C_{10}H_{16}$ Open chain Juniper berry (10%)	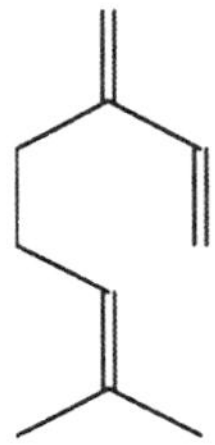
alpha-pinene $C_{10}H_{16}$ Bicyclic *Pinus sylvestris* (40%) (20%) Juniper berry (15%) *Eucalyptus globulus* (15%)	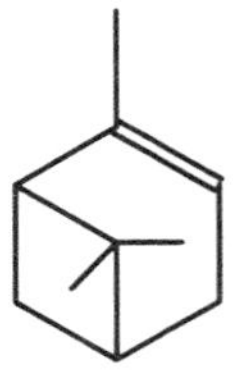
limonene $C_{10}H_{16}$ Monocyclic Sweet Orange (96%) Grapefruit (92%) Mandarin (70%) Lemon (65%) Black Pepper (20%)	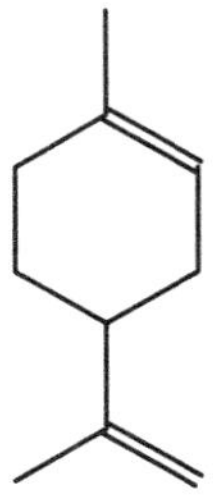

para-cymene $C_{10}H_{14}$ Aromatic *Thymus vulgaris* (35%) Oregano (15%) Sweet Marjoram (5%)	

A terpene with no double bonds is known as a saturated molecule, because all the carbon atoms are "saturated" with single bonds, and thus have little power to react without input of free energy. "Unsaturated" therefore means there is at least one C=C double bond. It appears that constituents which have several double bonds are more chemically reactive than constituents with no double bonds. One of the features of double bonds is they can undergo a reaction whereby one of the bonds is broken, and two other atoms or parts of molecules can be added to the two carbon atoms which were previously double bonded.

Naming - All monoterpenes have the ending **"ene"**. If you are looking at the name of a constituent, for example "myrcene", the -**ene** ending indicates the presence of at least one double bond between carbon atoms.

Solubility - Monoterpenes, as with most essential oil constituents, are not soluble in water, as they have long non-polar carbon chains. When a monoterpene has a polar functional group[13] added to it (for example a hydroxyl group), this makes the molecule (now a monoterpenol) more able to form emulsions where the oil can disperse in very tiny droplets. However, monoterpenes are soluble in other oils, such as almond oil or other vegetable oils, which are also non-polar. They are also fairly soluble in ethanol.

Volatility - This means the ease with which the constituent will evaporate. It is dependent on the boiling point of the constituent, which in turn is related to the structure of the constituent and its affinity for the other components in the essential oil. If a constituent is highly volatile, it may evaporate before it has a chance to penetrate the skin. Monoterpenes are among the most volatile of the groups of constituents, and are often

---

[13]See previous section for discussion of solubility and polarity.

thought of as "top" notes in perfumery, meaning that these are the odours that you first smell when you smell a blend.

Reactivity - Monoterpenes will combine over time with oxygen from the air, to form peroxides and thence epoxides and alcohols. It is thus important to keep them tightly sealed, and away from sources of free energy, such as heat and light, which speed up these decomposition reactions.

Toxic effects on the human body - Some monoterpenes are skin irritants, but it depends on the toughness of the skin and the reactivity of the particular monoterpene as to how severe the irritation will be. Sometimes the reaction is more in the form of a sensitisation of the skin, with repeated use of oils containing high percentages of monoterpenes. This effect has been reported anecdotally by a few aromatherapists I know, and is supported by Tisserand and Balacs (p.199, 1995)[14]. They say that the "allergic effect of terpene-rich oils [is] possibly due to hydroperoxide formation on storage." Terpene-rich oils would be all the citrus oils, Pine, Juniper, Black Pepper and Cypress oils, and to a lesser extent Rosemary, Tea tree and Eucalyptus.

Hausen et al. (1999)[15] researched the sensitising effects of tea tree oil and the monoterpenes it contains by patch-testing 15 human volunteers, and found that oxidised tea tree oil, in particular the peroxides, epoxides and endoperoxides that were formed from the monoterpenes in the oil were the sensitising agents. Para-cymene was also found to increase in the oxidised product, at the expense of alpha- and gamma-terpinenes.

Alpha- and beta-pinene and delta-3-carene, which are constituents of turpentine oil have been investigated as likely respiratory irritants in saw mills in Sweden. The uptake of the constituents from the air was between 62-68%, and after 2 hours exposure, 2-5% was excreted unchanged in expired air. The mean half lives of the constituents in the body were 32, 25 and 42 hours respectively for the test amount of 450mg/m3 of turpentine in the air. At the end of the exposure subjects experienced discomfort in the throat and airways, and airway resistance was increased

---

[14] Tisserand R. & Balacs T, "Essential Oil Safety" (1995), Churchill Livingstone, Edinburgh.

[15] Hausen B M, Reichling J, Harkenthal M, "Degradation products of monoterpenes are the sensitizing agents in tea tree oil" *American Journal of Contact Dermatitis* 10(2) pp. 68-77.

after the end of the exposure (Filipsson, 1996)[16]. Depending on the administration method, inhalations of pinene-rich oils may reach this concentration, so it is as well to ask whether people suffer from asthma or other conditions which are already increasing their airway resistance.

Dr. Pénoël suggests that Turpentine and Juniper twig essential oils give a "definite nephrotoxicity" (p. 222), which may be due to sabinyl acetate content (an ester showing marked toxicity, though not necessarily nephrotoxicity, Tisserand & Balacs, 1995 p.197). However, Schilcher and Leuschner (1997)[17] tested for possible kidney damage in rats which were fed up to 1000 mg of Juniper oil, and found no functional or morphological changes.

Therapeutic effects of Monoterpenes

Dr Daniel Pénoël has worked as a doctor and aromatherapist in France for over twenty years. Together with a chemist, Pierre Franchomme, he wrote a comprehensive book "L'aromathérapie exactement", which summarised the great bulk of research into essential oils prior to 1990[18]. I have used his book as a framework, because he conveniently categorises the therapeutic effects of essential oils by functional group. In each section, I list some of the information which he has found on the medicinal properties of the specific functional groups of constituents. Where possible, I supplement his observations with source papers from medical journals (found by searching the Medline database on the Internet). These are the properties proposed for monoterpenes:

- **General tonic and stimulant** - particularly for mucous glands, which are stimulated either to dry up, or to produce a more fluid mucus. The administration method would be inhalation. The net effect is **decongestant**. The pinenes are especially noted by Dr. Pénoël as mucolytics (Pénoël & Franchomme, 1990, p. 219), which is interesting in light of the airway irritation mentioned above. However, I have not been able to find supportive evidence in the research literature for the mucolytic effects of the pinenes.

---

[16] Filipsson AF, (1996) "Short term inhalation exposure to turpentine: toxicokinetics and acute effects in men" *Occupational and Environmental Medicine* 53(2) p. 100-5

[17] Schilcher H & Leuschner F (1997) "The potential nephrotoxic effects of essential juniper oil" *Arzneimittelforschung* July 47(7) pp. 855-8.

[18] Pénoël D & Franchomme P (1990) "L'aromathérapie exactement", Roger Jollois, Limoges

- Limonene has been used as an agent to dissolve **gall stones**, when directly injected into the biliary system. It dissolved the gall stones completely in 48% of cases in a study by Igimi et al. (1991)[19], and subsequent research indicates that it is most efficacious where the gallstones are cholesterol based Igimi et al. (1992)[20]. It is doubtful that the amounts used in a typical aromatherapy treatment would have any impact on gallstones, but it is worth a try. Limonene is found in a great many essential oils, and in high percentages in the citrus peel oils.

- **Cancer prevention and treatment**. This is an area receiving a great deal of research attention, and it has been found that several essential oil constituents show promise as cancer prevention agents. They prevent the initiation, promotion and progression of cancers, and show promise as therapeutic agents. Limonene in particular has been investigated, and the monoterpenol perillyl alcohol. See Chapter 7 for the structures.

The cancers which have been treated in rats are breast and pancreatic carcinomas. Limonene and perillyl alcohol appear to inhibit the isoprenylation of small G proteins, thus altering the regulation of certain genes, the products of which are responsible for blocking cancer cells' metabolic processes (Gould, 1997)[21]. Clinical trials with humans are underway, but no conclusive results are available at the time of printing.

- **Hormone-like properties** - certain of the monoterpenes from the *Pinus* genus supposedly act on the pituitary-adrenal system, causing stimulation of adrenaline production (Pénoël & Franchomme, 1990, p.219).

---

[19] Igimi H, Tamura R, Yamamoto F, Toraishi K, Kataoka A, Ikejiri Y, Hisatsugu T, Shimura H (1991) "Medical dissolution of gallstones. Clinical experience of d-limonene as a simple, safe, and effective solvent" *Digestive Diseases And Sciences* 36(2) pp. 200-8
[20] Igimi H, Watanabe D, Yamamoto F, Asakawa S, Toraishi K, Shimura H (1992) "A useful cholesterol solvent for the medical dissolution of gallstones" *Gastroenterologica Japonica* 27(4) pp. 536-45.
[21] Gould MN, (1997) "Cancer chemoprevention and therapy by monoterpenes" *Environmental Health Perspectives* June, v. 105 Supplement 4, pp. 977-979.

- **Muscular aches relieved** - Para-cymene, found in several oils, is also used as a chemical in pharmaceutical preparations that are specified for muscular aches (Pénoël & Franchomme, 1990, p.219).

Essential oils with high percentages of monoterpenes

Angelica (root) *Angelica archangelica*	alpha-pinene 25%	1,8-cineole 14.5%	alpha-phellandrene 13.5%
Elemi *Canarium luzonicum*	limonene 54%	alpha-phellandrene 15.1%	elemol 15%
Grapefruit (Israel)	limonene 93%	myrcene 1.97%	alpha-pinene 0.59%
Lemon (Argentina)	limonene 70%	beta-pinene 11%	gamma-terpinene 8%
Lime (Persian)	limonene 58%	gamma-terpinene 16%	beta-pinene 6%
Mandarin (Italy)	limonene 71%	gamma-terpinene 18.54%	alpha-pinene 2.39%
Orange Sweet (Brazil)	limonene 89%	myrcene 1.71%	beta-bisabolene 1.29%
Fir balsam (Canada) *Abies balsamea*	beta-pinene 30%	delta-3-carene 21.45%	bornyl acetate 11.85%
Juniper berry	alpha-pinene 33%	myrcene 11%	beta-farnesene 10.5%
Nutmeg	alpha-pinene 22%	sabinene 18.55%	beta-pinene 15.55%
Frankincense (Somalia)	alpha-pinene 34.53%	alpha-phellandrene 14.6%	para-cymene 14%
Dwarf Pine *Pinus mugo ssp. pumilia*	delta-3-carene 35%	beta-phellandrene 15%	alpha-pinene 13.1%
Pine *Pinus pinaster*	alpha-pinene 44.1%	beta-pinene 29.5%	myrcene 4.69%
Pine *Pinus sylvestris*	alpha-pinene 42%	delta-3-carene 20.5%	limonene 5.2%
Bergamot (Calabria)	limonene 38.4%	linalyl acetate 28%	linalool 8%
Rosemary (Spain) (? camphor chemotype)	alpha-pinene 22%	camphor 17%	1,8-cineole 17%

SESQUITERPENES

Structure - Sesquiterpenes are hydrocarbon molecules which have 15 carbon atoms (three isoprene units), and a varying number of hydrogen atoms. There are no oxygen atoms. In Latin, "sesqui-" means "one and a half".

Examples

beta-caryophyllene $C_{15}H_{22}$ Bicyclic Black Pepper (30%) Clove Bud (10%)	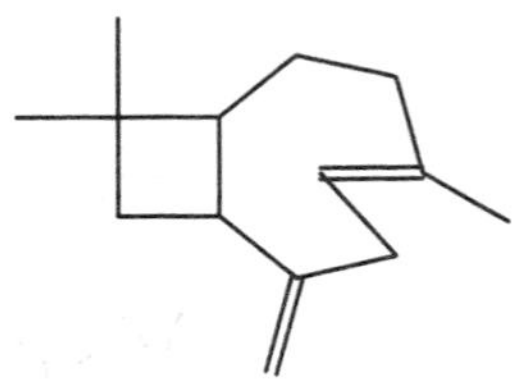
chamazulene $C_{14}H_{16}$ Bicyclic (not quite sesquiterpenoid - only 14 C atoms) German Chamomile (15%)	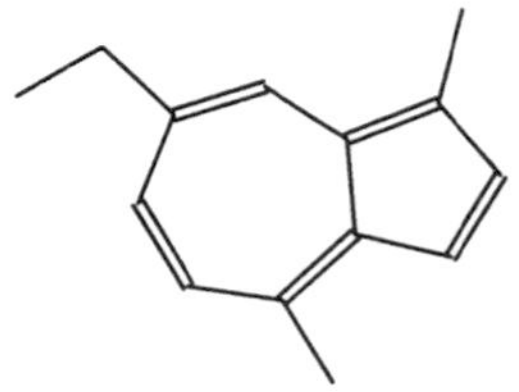
beta-farnesene $C_{15}H_{22}$ Open chain Ginger (2%)	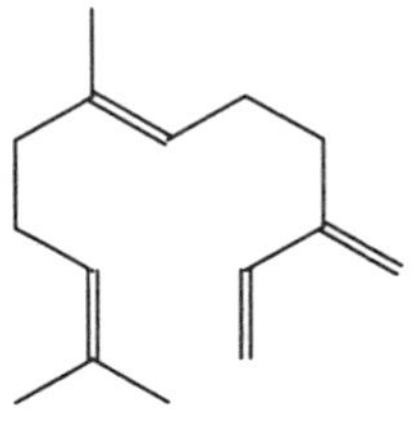

Naming - Being terpenes, they also end "**-ene**", and like the monoterpenes, the rest of the name is derived either from the type of plant it was found in, or from the country it was found in. For example, alpha-pinene was first found in pine oil (*Pinus sylvestris*), and alpha-humulene was first found in oil of hops (*Humulus lupulus*). Apart from beta-caryophyllene, most of the sesquiterpenes occur in only a few different oils, for example chamazulene in German Chamomile oil, cedrene in the Cedarwood oils and santalene only in Sandalwood oil.

Solubility - Sesquiterpenes are not soluble in water, though they do dissolve readily in other oils.

Volatility - Due to their higher boiling points, sesquiterpenes are not so volatile as monoterpenes, and this also accounts for their presence as middle or base notes in perfumery. They feel more viscous than the monoterpenes, and an oil high in sesquiterpenes will pour more slowly out of the bottle.

Reactivity - Similarly to monoterpenes, sesquiterpenes will react over time with oxygen in the air, and if exposed to free energy in the form of heat or light, will form epoxides and alcohols and will eventually polymerise to form long chain resins. In some oils, such as Patchouli and Vetiver, this oxidation is thought to actually improve the odour, and as such, the aged oils are often prized above the newly distilled ones. As far as comparison of the therapeutic effects between new and old oils, I have not seen any research as yet.

Toxic effects on the body - Sesquiterpenes may also become skin irritants when oxidised, but generally seem to not be as irritating as the monoterpenes can be.

Therapeutic effects of Sesquiterpenes

- **Anti-inflammatory** - Sesquiterpenes with several double bonds are supposed to be good for reducing inflammation caused by stings and bites, and for histaminic reactions (Pénoël (1990), p.220)

  Beta-caryophyllene has been found to reduce stomach cell damage from alcohol poisoning in rats when administered orally. It also seems to work in a better way than other non-steroidal anti-inflammatory preparations as it doesn't damage the gastric mucosa (Tambe et al, 1996)[22].

  Chamazulene, found in German Chamomile oil has been shown to be anti-inflammatory *in vivo*, and Safayhi et al. (1994)[23] found by *in vitro* studies that the anti-inflammatory effect is due to the blocking of formation of leukotriene B4 in neutrophils. Leukotriene B4 is one

---

[22] Tambe Y, Tsujiuchi H, Honda G, Ikeshiro Y, Tanaka S (1996) " Gastric cytoprotection of the non-steroidal anti-inflammatory sesquiterpene, beta-caryophyllene." *Planta Medica,* 62(5) pp. 469-70

[23] Safayhi H, Sabieraj J, Sailer ER, Ammon HP. (1994) "Chamazulene: an antioxidant-type inhibitor of leukotriene B4 formation." *Planta Medica* 60(5) pp. 410-3.

of the chemicals released at the site of inflammation, which causes further inflammation.

- **Possible hormonal effects** - Farnesene is a molecule which occurs throughout the plant world, and also curiously enough in a form known as "farnesyl pyrophosphate" in humans and other animals. In humans it is part of the chemical pathway to the formation of cholesterol, and the steroid hormones[24]. In mice, as Ma et al. (1999) found out, farnesene is present in the secretions of female mice which indicate the onset of estrus, and is responsible for inducing groups of female mice to go into estrus together[25]. Further research may indicate that farnesene is also present in human female secretions, and may perform the same function. However, this is just my wild guess, as the research has not been done.

Essential oils with high percentages of sesquiterpenes

Myrrh gum *Commiphora myrrha* (headspace)	delta-elemene 28.7%	alpha-copaene 10%	beta-elemene 6.1%
Cedarwood Texas *Juniperus mexicana*	thujopsene 32%	alpha-cedrene 24.1%	cedrol 16%
Cedarwood Virginia *Juniperus virginiana*	cedrol 26%	alpha-cedrene 24.5%	thujopsene 15%
German Chamomile (Bulgaria) *Chamomilla recutita*	farnesene 27%	chamazulene 17%	alpha-bisabolol oxide B 11%
Ginger (China) *Zingiber officinale* Roscoe	ar-curcumene 16.3%	alpha-zingiberene 14.2%	beta-sesquiphellandrene 10.6%

---

[24] Lehninger A (1982) "*Principles of Biochemistry*" Worth Publishers, NY, p.608

[25] Ma W, Miao Z, Novotny MV (1999) "Induction of estrus in grouped female mice (Mus domesticus) by synthetic analogues of preputial gland constituents." *Chemical Senses*. 24(3) pp. 289-93.

CHAPTER FOUR

FUNCTIONAL GROUPS

A "functional group" is a small part of a molecule, made up of two or more atoms, which gives the whole molecule different properties. Functional groups can make a molecule more polar, more reactive, or more susceptible to rearrangement of structure. They also affect the therapeutic properties of molecules.

In essential oils, functional groups are made up of different combinations of oxygen and hydrogen atoms, and are attached to the basic terpenoid skeleton. As the functional group(s) of a molecule give it distinct properties, we can group molecules by their different functional groups, and predict that members of the same group will have similar chemical properties. Generalisations can be made with caution about the therapeutic properties of molecules based on their structure, but research is needed to discover the specific actions of each different constituent, alone and in synergy with the other constituents in the oil.

ALCOHOLS

Terpenoid alcohols have an hydroxyl group attached to one of their carbon atoms. An hydroxyl group is also known as an -OH group, because it is exactly that: an oxygen atom, bonded on the one hand to a carbon atom in the chain, and on the other hand, to a hydrogen atom. Phenols are also a type of alcohol, but due to their aromatic ring, they are classed differently from the other types of alcohols. We will come to them later. To tell whether a diagram of a molecule is an alcohol, you look for an -OH group joined onto a carbon atom.

MONOTERPENOLS

Structure - They all have the monoterpenoid carbon chain of 10 carbon atoms, and may contain an hydroxyl group anywhere along the chain. The addition of an -OH group usually means the number of double bonds reduces by one.

Examples

linalol (or linalool) $C_{10}H_{17}OH$ Open chain A tertiary alcohol Ho Leaf & Rosewood (90%) Basil (linalool) (40%) Lavender (37%)	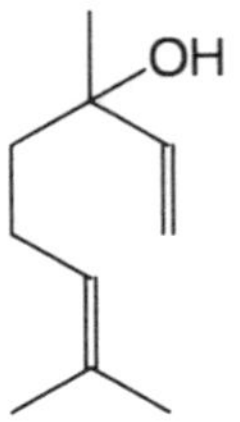
terpinen-4-ol $C_{10}H_{16}OH$ Monocyclic A tertiary alcohol Tea tree (40%) Sweet Marjoram (25%) Nutmeg (6%)	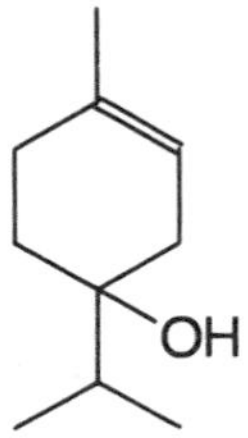
menthol $C_{10}H_{19}OH$ A secondary alcohol Cornmint (65%) Peppermint (45%)	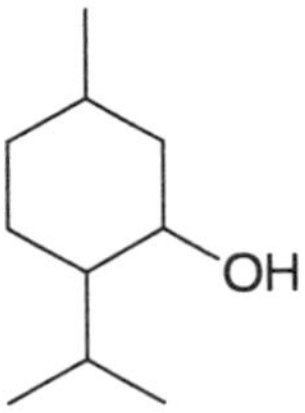
geraniol $C_{10}H_{17}OH$ A primary alcohol Palmarosa (75%) Citronella (25%) Geranium (20%)	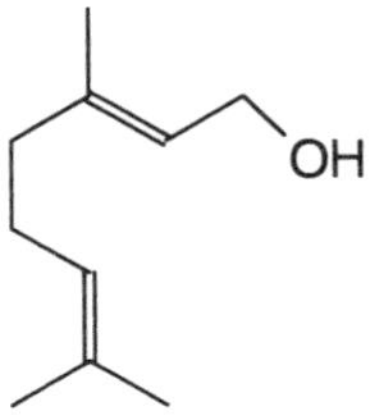

The -OH group can be added to carbon atoms at different places in the molecule, yielding primary, secondary and tertiary alcohols. A primary alcohol is where the -OH group is attached to a carbon atom which in turn is attached to only one other carbon atom. A secondary alcohol has two carbon atoms attached to the -OH carbon, and a tertiary alcohol has three carbon atoms attached to that same carbon atom. This has relevance in terms of how easily the alcohol can be oxidised, which in turn affects its fate in the body in terms of how the liver will deal with it. It also has implications for the speed of degradation of the oil over time.

Primary alcohols are easily oxidised to aldehydes, secondary alcohols are oxidised to ketones, and tertiary alcohols are not oxidised at all.

Naming - Alcohols will all have the ending " **-ol**". The rest of their name will come from their parent monoterpene. An example is terpinen-4-**ol** which is derived from terpinene. The "4" indicates the position on the molecule where the -OH group is attached, because you count round the ring starting from the top, keeping the numbers as low as possible.

Solubility - Monoterpenols are most often soluble in ethanol, and will also be soluble in other oils, due to their long carbon chain skeleton. Due to their polar group, they will have slight solubility in water (0.1 - 0.4 g/L).[26]

Volatility - Monoterpenols are often less volatile than monoterpenes, depending on their boiling points, and analogous structures such as terpineol and limonene show the monoterpenol to be more viscous and less volatile.

Reactivity - When exposed to the air, monoterpenols can oxidise to aldehydes and acids, or form resins, which means that they should also be stored away from the air and sources of free energy such as light and heat.

Toxic effects on the body - Monoterpenols are not thought to exhibit any significant toxicity (Tisserand & Balacs, 1995, p.182)[27] unless taken orally in amounts exceeding several mls of oil. Some can be mildly irritating to the skin, but no more than monoterpenes.

Therapeutic effects of monoterpenols

- **Anti-infectious** - Monoterpenols have been shown to have strong anti-bacterial and anti-fungal properties, which, though not as strong as those of commercial disinfectants, have the advantage that they are usually mild on the skin and mucous membranes (Pénoël & Franchomme, 1990, pp.116-165). Pattnaik et al. (1997)[28] investigated the activity of linalool, geraniol and menthol against bacteria and

---

[26.] These details are from Guenther E, (1972) "The Essential Oils" Krieger, Malabar, Flanders.

[27] Tisserand R & Balacs T, (1995) "Essential Oil Safety" Churchill Livingstone, London , p. 182

[28] Pattnaik S, Subramanayam VR, Bapaji M, Kole CR (1997) "Antibacterial and antifungal activity of aromatic constituents of essential oils" *Microbios* 89(358) pp. 39-46

fungi, and found that linalool was best against bacteria, and geraniol was best against fungi, whereas menthol only inhibited half of the eighteen bacteria and twelve fungi tested. Carson & Riley (1995)[29] found that terpinen-4-ol, linalool and alpha-terpineol were active against *Escherichia coli*, *Staphylococcus aureus*, and *Candida albicans*. Terpinen-4-ol was the only constituent tested which was active against *Pseudomonas aeruginosa*. Budhiraja et al. (1999)[30] suggest that terpinen-4-ol, the major constituent of tea tree (*Melaleuca alternifolia*) oil activates monocytes (white blood cells which deal with infection).

- **Vaso-constrictive properties** - Menthol, linalool, alpha-terpineol and geraniol, make the site of application feel cold. They also induce a slight local anaesthesia at the site of application (Pénoël & Franchomme, 1990, p 158).

- **Tonifying and stimulant properties** - Some monoterpenols have specific systems which they stimulate, for example terpinen-4-ol. *Origanum marjorana* contains about 22% of terpinen-4-ol, and about 50% of monoterpenols all told. But most times Sweet Marjoram is thought of as a calming oil. The reason for this (according to Pénoël & Franchomme, 1990, p. 158) is that the exhausted client receiving the marjoram treatment will in fact have their exhausted psycho-neuro-endocrino-immune systems boosted by the high alcohol content and the relief to the body at being replenished is such that a deep relaxation ensues.

- **Sedative properties** - One monoterpenol, linalool, exhibits some interesting sedative properties, even through inhalation. Buchbauer et al. (1991) showed that mice which had either lavender oil (which contains about 40% linalool) and both linalool and its ester linalyl acetate pumped into their cage, all experienced a marked reduction in movement compared to the controls[31].

---

[29]Carson CF, Riley TV. (1995) " Antimicrobial activity of the major components of the essential oil of Melaleuca alternifolia" *Journal of Applied Bacteriology* 78(3) pp. 264-9.
[30] Budhiraja SS, Cullum ME, Sioutis SS, Evangelista L, Habanova ST (1999) "Biological activity of Melaleuca alternifola (Tea Tree) oil component, terpinen-4-ol, in human myelocytic cell line HL-60." *Journal of Manipulative Physiological Therapeutics* 22(7) pp. 47-53
[31] Buchbauer G, Jirovetz L, Jager W, Dietrich H, Plank C (1991) "Aromatherapy:

Elisabetsky et al. (1999)[32] examined the effect of linalool on the binding of L-[H3]-glutamate to neuronal membranes in the central nervous system. L-[H3]-glutamate is an excitatory neurotransmitter associated with convulsions. The experiment showed that linalool was an antagonist for glutamate *in vitro*, and delayed aspartate-induced convulsions and blocked quinolinic acid-induced convulsions. This means that linalool may work as a sedative at the level of the central nervous system by modifying the response of neurons to L-[H3]-glutamate.

Lis-Balchin & Hart (1999)[33] found that linalool and lavender oil seemed to exert a spasmolytic effect on guinea pig intestinal smooth muscle. They suggest that the mechanism is through cyclic AMP modulation, which affects the ability of smooth muscle to contract.

Essential oils with high percentages of monoterpenols

Basil (Portugal)	linalool 38.2%	methyl chavicol 16.4%	beta-caryophyllene 7%
Rosewood *Aniba rosaeodora*	linalool 85.3%	alpha-terpineol 3.5%	cis-linalool oxide 1.5%
Citronella (Ceylon) *Cymbopogon nardus* L. Rendle	geraniol 18%	limonene 9.7%	citronellol 8.4%
Geranium Bourbon	citronellol 21.2%	geraniol 17.45%	linalool 12.9%
Ho leaf *Cinnamomum camphora* Sieb. *ssp.formosana* var. *brientalis*	linalool 95%	camphor 0.4%	limonene 0.2%
Lavandin (Abrial quality, France) *Lavandula hybrida* Rev.	linalool 33.5%	linalyl acetate 27.1%	camphor 9.5%
Sweet Marjoram *Marjorana hortensis* Moench (*Origanum majorana*)	terpinen-4-ol 36.3%	cis-sabinene hydrate 15.9%	para-cymene 9.5%

evidence for sedative effects of the essential oil of lavender after inhalation." *Zeitschrift fur Naturforschung [C]*. Nov-Dec. 46(11-12) pp. 1067-72

[32] Elisabetsky E, Brum LF, Souza DO (1999) "Anticonvulsant properties of linalool in glutamate-related seizure models" *Phtyomedicine* 6(2) pp. 107-13

[33] Lis-Balchin M, Hart S (1999) "Studies on the mode of action of the essential oil of Lavender (Lavandula angustifolia P. Miller)" *Phytotherapy Research* 13(6) pp. 540-2.

Neroli bigarade	linalool 37.5%	limonene 16.6%	beta-pinene 11.8%
Palmarosa (India) *Cymbopogon martinii* Stapf. var. *motia*	geraniol 80%	geranyl acetate 8.25%	linalool 2.79%
Peppermint (USA) *Mentha piperita* L. var. Mitcham	menthol 42.8%	menthone 19.4%	sabinene hydrate 6.6%
Rose (Bulgaria) *Rosa damascena* Mill. (otto)	citronellol 33.4%	stearopten waxes 24%	geraniol 18%
Rose (Egypt) *Rosa damascena* Mill.	2-phenyl ethyl alcohol 37.9% *possibly adulterated?	geraniol 15.8%	citronellol 12.6%
Tea Tree *Melaleuca alternifolia*	terpinen-4-ol 45.4%	gamma-terpinene 15.7%	alpha-terpinene 7.1%

SESQUITERPENOLS

Structure - The sesquiterpenols are derived from sesquiterpenes, though often you do not get both in the same oil. It seems that different plants create special sesquiterpenols which become characteristic for that plant. Examples are patchoulol, which is only found in Patchouli oil, and the santalols found in Sandalwood oil.

Naming - Also end in "-ol", sometimes deriving the first part of their name from parent sesquiterpenes (e.g. farnesene and farnesol).

Solubility - Soluble in alcohol and organic oils. Not soluble in water, in spite of the -OH group, because of the long carbon chain.

Examples

farnesol $C_{15}H_{25}OH$ Open chain unsaturated Jasmine sometimes (10%) Ylang ylang (2%) *Rosa damascena* (1%)	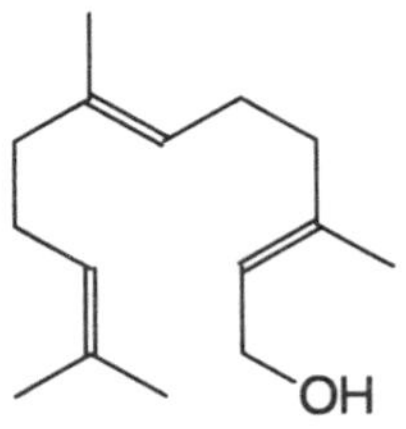

viridiflorol $C_{14}H_{23}OH$ Tricyclic saturated Niaouli (18%)	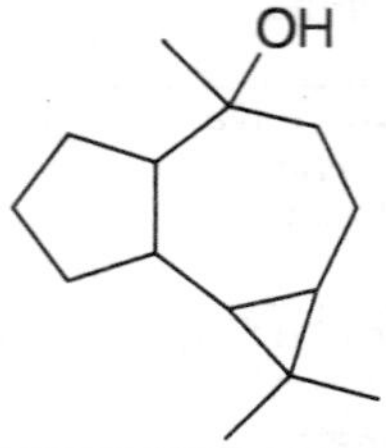
patchoulol $C_{15}H_{24}OH$ Tricyclic saturated Patchouli (40%)	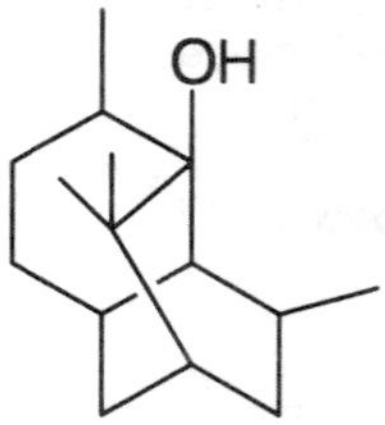
beta-santalol $C_{15}H_{24}OH$ Bicyclic unsaturated Sandalwood (20%)	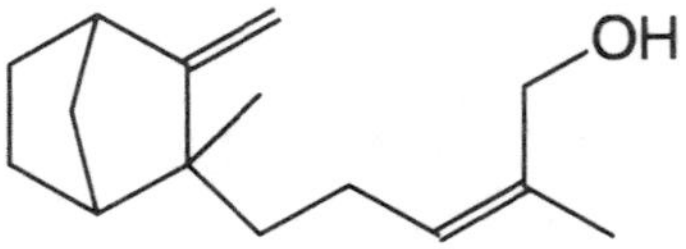

Volatility - Again, this depends on boiling point, which in turn is dependent on structure, and internal attractions of the molecules for each other. The greater the attraction, the higher the boiling point and the less volatile they will be. For example, the two stereoisomers of farnesol have different boiling points. The one with the -OH group tucked inside the molecule is less likely to be attracted to other molecules, and therefore it has a lower boiling point than the one with the -OH group sticking out.

Reactivity - Similar to monoterpenols, in that they will oxidise in air, thus possibly losing their therapeutic properties. They may gain different odour characteristics too, which may be desired by the perfumery trade.

Toxic effects on the body - Sesquiterpenols seem to be relatively harmless, and have even lower levels of irritation than monoterpenols.

Therapeutic effects of sesquiterpenols
- **General tonic** - This is not so great an effect as with the monoterpenols, and again, specific sesquiterpenols will have specific functions. Dr. Pénoël suggests that: viridiflorol from Niaouli oil may have an oestrogenic effect and a vein tonic effect; cedrol from

Juniperus virginana oil may have a vein tonic effect; and that santalol from Sandalwood oil is a cardiotonic (Pénoël & Franchomme, 1990, p.117). Unfortunately, he has not specifically referenced these suggestions so we can't pursue them further.

- **Anti-inflammatory** - Alpha-bisabolol is a sesquiterpenol found in German Chamomile oil, and is even more anti-inflammatory than chamazulene[34]. It has been used in German pharmaceutical preparations for cracked nipples. Several other sesquiterpenol-rich oils show anti-inflammatory properties, although I could not find the research to show that it is actually the sesquiterpenols causing the effect.

- **Possible vascular effects** - Luft et al. (1999)[35] researched the effects of farnesol on smooth muscle cells of the rat aorta. They noted that farnesol blocks L-type Ca2+ channels, thereby preventing contraction. When they fed farnesol to rats with hypertension (0.5g/kg doses), farnesol significantly reduced the experimental rats' blood pressure for up to 48 hours. It may be that dietary farnesol would have similar effects in humans, though topically applied farnesol is unlikely to have significant effect, as it may not even penetrate the epidermis due its molecular size. Cadinol isomers also appeared to relax K+ induced contractions of rat aortas, as Zygmunt et al. noted (1993)[36].

- **Neuronal effects** - Alpha-eudesmol alleviates the over-stimulation of neuronal cells by blocking Ca2+ channels in neurons in the brain. Found in *Juniperus virginiana* (Cedarwood Virginian) oil, alpha-eudesmol particularly blocks those Ca2+ channels which are responsible for releasing glutamate during artifically induced ischemic strokes in rats. Alpha-eudesmol was introduced into the intracerebroventricular fluid, and significantly reduced the amount of

---

[34] Carle R, Gomaa K (1992) "The medicinal use of Matricaria flos" *British Journal of Phytotherapy* 2(4) pp. 147-153

[35] Luft UC, Bychkov R, Gollasch M, Gross V, Roullet JB, McCarron DA, Ried C, Hofmann F, Yagil Y, Yagil C, Haller H, Luft FC (1999) "Farnesol blocks the L-type Ca2+ channel by targeting the alpha 1C subunit" *Arteriosclerosis Thrombosis and Vascular Biology.* Apr 19(4) pp. 959-66.

[36] Zygmunt PM, Larsson B, Sterner O, Vinge E, Hogestatt ED, (1993) "Calcium antagonistic properties of the sesquiterpene T-cadinol and Calcium antagonistic properties of the sesquiterpene T-cadinol and related substances: structure-activity studies" *Pharmacology & Toxicology* 73(1) pp.3-9.

brain damage and water volume in the brain after the brain injury (Asakura et al. 2000)[37].

Beta-eudesmol, found in *Amyris balsamifera* (West Indian Sandalwood) oil, alleviated electroshock convulsions in mice. It also reduced high level potassium induced seizures and prevented the toxicity of organophosphate nerve poisoning. It may have potential as an antiepileptic substance, either on its own, or in conjunction with phenytoin, an antiepileptic drug (Chiou et al, 1997)[38]. It is also found in some Chinese herbs, and also in small amounts in Ginger and *Helichrysum italicum* (Everlasting or Imortelle) oils.

- **Anti-cancer effects** - Rioja et al. (2000)[39] examined the effects of farnesol on affected blood cells from patients with acute myeloid leukemia. It seemed that farnesol selectively kills leukaemic cells in preference to normal haemopoietic cells (cells which blood cells are derived from), at a level of 30 microM of farnesol. Whether there will be similar effects *in vivo* remains to be seen. There are several isomers of farnesol, and it wasn't specified which isomer was used in the study. However, the oils that any type of farnesol is found in are *Helichrysum odoratissimum* (16%), Cananga oil (21%), sometimes in Jasmine (12%) and Lemongrass (9%). It is also found in smaller amounts in Ylang ylang oil and *Rosa damascena* oil.

Dietary nerolidol inhibited the growth of artificially induced neoplasms in the large intestine of male rats, and reduced the numbers of tumours present at the end of the test period. Whether similar effects would be found in humans is not known, but Wattenberg (1991)[40] suggests that nerolidol be investigated as an inhibitor of large

---

[37] Asakura K, Matsuo Y, Oshima T, Kihara T, Minagawa K, Araki Y, Kagawa K, Kanemasa T, Ninomiya M. (2000) " omega-Agatoxin IVA-sensitive Ca(2+) channel blocker, alpha-eudesmol, protects against brain injury after focal ischemia in rats." *European Journal of Pharmacology* 394(1) pp.57-65.

[38] Chiou LC, Ling JY, Chang CC, (1997) " Chinese herb constituent beta-eudesmol alleviated the electroshock seizures in mice and electrographic seizures in rat hippocampal slices." *Neuroscience Letters*. August 15, 231(3) pp. 171-4.

[39] Rioja A, Pizzey AR, Marson CM, Thomas NS. (2000) "Preferential induction of apoptosis of leukaemic cells by farnesol." FEBS Letters.Feb 11; 467(2-3) pp.291-5.
[40] Wattenberg LW (1991) "Inhibition of azoxymethane-induced neoplasia of the large bowel by 3-hydroxy-3,7,11-trimethyl-1,6,10-dodecatriene (nerolidol)." *Carcinogenesis* Jan 12(1) pp. 151-2.

intestine carcinogenesis. The best source of nerolidol (about 90%) is found in a chemotype of *Melaleuca quinquinervia* oil[41] - the other chemotype is the 1,8-cineole chemotype, which is commonly known as Niaouli oil. Several of the spice oils such as Cardamom and Ambrette seed contain small amounts of nerolidol.

- **Anti-viral activity** - Sandalwood oil is nearly all sesquiterpenes and sesquiterpenols, santalenes and santalols. Benencia and Courreges (1999)[42] tested it against *Herpes simplex* viruses I and II *in vitro*, and found that the oil inhibited replication of both viruses, but did not kill them. It was less effective at high levels of viral numbers. This may mean that sandalwood oil could be used as an early on-set preventative for cold sores.

- **Possible anti-malarial activity** - Lopes et al (1999)[43] found that nerolidol (see above under anti-cancer effects) completely inhibited the development of the malaria protozoan *in vitro*, seemingly by inhibition of glycoprotein synthesis. There are other types of constituents which show definite anti-malarial activity, such as some of the *Artemisia sp.* lactones, but I thought I would put this in, as the Amazon Waiapi Indians treat malaria by inhaling the vapor from a forest tree (*Viola surinamensis*) which contains nerolidol.

Essential oils with high percentages of sesquiterpenols

Cedarwood Virginia *Juniperus virginiana*	cedrol 26%	alpha-cedrene 24.5%	thujopsene 15%
Patchouli (Indonesia) *Pogostemon cablin* Benth.	patchouli alcohol 33%	alpha-patchoulene 22%	beta-caryophyllene 20%
Sandalwood (India)	cis-alpha-santalol 50%	cis-beta-santalol 20.9%	epi-beta-santalol 4.1%
Vetiver *Vetivera zizanoides* Stapf.	vetiverol 50%	vetivenes 20%	alpha-vetivol 10%

---

[41] Lassak EV and Southwell IA (1977) "Essential Oil Isolates from the Australian Flora" *IFFA* May/June, pp. 126-132.
[42] Benencia F, Courreges MC (1999) "Antiviral activity of sandalwood oil against herpes simplex viruses-1 and -2´ *Phytomedicine*. May;6(2) pp. 119-23
[43] Lopes NP, Kato MJ, Andrade EH, Maia JG, Yoshida M, Planchart AR, Katzin AM. (1999) "Antimalarial use of volatile oil from leaves of Virola surinamensis (Rol.) Warb. by Waiapi Amazon Indians." *Journal of Ethnopharmacology* Nov 30; 67(3) pp. 313-9.

PHENOLS

Structure - Phenols have an hydroxyl (-OH) group, attached to an aromatic or benzene ring. This is what distinguishes them from alcohols. The benzene ring is represented with a circle in the middle of the six-membered ring because all the bonds in the ring between carbon atoms have equal value, of 1.5 bonds. Sometimes it is drawn as a ring with three double bonds, but the circle representation is more accurate.

The aromatic ring has the effect of amplifying the electronegativity of the oxygen atom, so the hydrogen atom of the hydroxyl group is quite positive. The result of this is that phenols begin to react a little like acids, which are characterised by the ease with which they can give up their hydrogen atoms to form ions. In essential oils, the phenols usually have a three carbon chain attached to them, which makes them into phenyl propanoid compounds. ("Propan-" is the chemistry jargon stem for "three carbon chain").

Examples

thymol $C_{10}H_{12}OH$ *Thymus vulgaris* (40%)	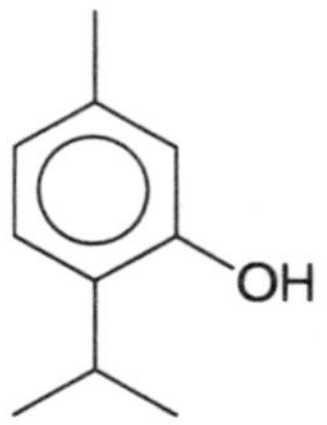
carvacrol $C_{10}H_{12}OH$ *Origanum vulgaris* (60%)	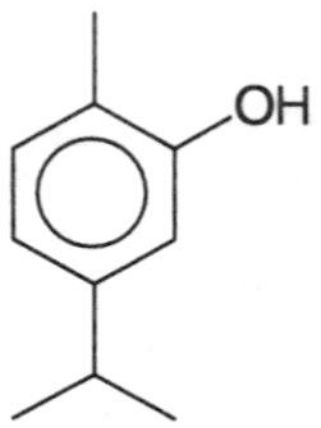
eugenol $C_9H_8OCH_3OH$ *Pimento dioica* Allspice (80%) Clove bud (70%)	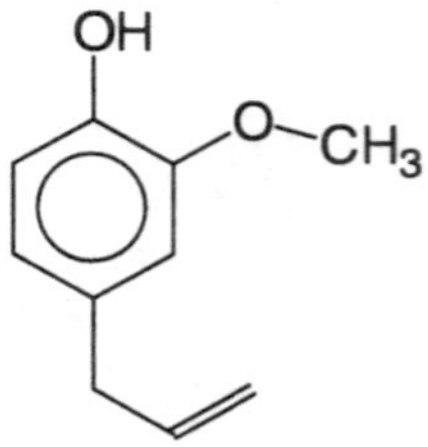

chavicol C₉H₉OH *Pimenta racemosa* West Indian Bay (20%)	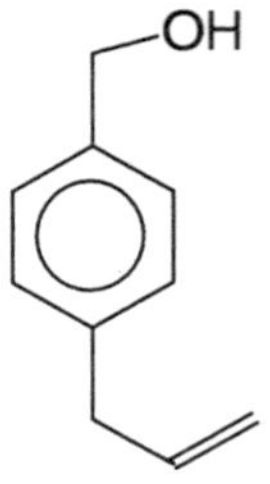

Naming - Because phenols contain an -OH group, they also have names ending **"-ol"**. Fortunately, the four described above are the most common phenols in essential oils, so nearly all other "-ol" ending words will be ordinary alcohols.

Solubility - Due to the aromatic ring, which can disperse the excess negative charges of the oxygen atom, the hydrogen atom of the -OH group becomes very positive, and this makes phenols into polar molecules, which are, like the alcohols, more soluble in water than hydrocarbons. Phenols are soluble up to 1 g/L. For aromatherapy purposes, this still means that they are not very soluble in water, but they do dissolve in ethanol, and other oils.

Volatility - Due to internal bonding via their hydroxyl (-OH) group, phenols often are found as crystals (e.g. thymol) at room temperature, and they have boiling points above 200 C. They thus do not evaporate very quickly, and have a greater chance of penetrating the skin.

Reactivity - Phenols are known as reactive molecules in chemistry, eliciting "hazardous chemical" labels. They will also readily release their hydroxyl hydrogen atom, and bind with other polar or positively charged molecules, including the protein molecules of the skin, thereby causing damage to the normal structure of the skin.

Toxic effects on the body - Phenols are the most irritant of the constituents to the skin and mucous membranes, and can cause contact dermatitis, and also sensitisation dermatitis. In order to excrete them, the liver has to convert them into sulphonates, and large internal doses could cause damage to the liver cells by depleting them of sulfonates (Pénoël & Franchomme, 1990, p.163).

Therapeutic effects of phenols

- **Tonics and general stimulants** - Phenols are the most stimulating of the constituents, perhaps too stimulating, and should only be used internally with extreme care.

- **Anti-infectious agents** - Phenols prevent most micro-organisms including *Staphylococcus aureus* and *Pseudomonas aeruginosa* from growing, and in most cases kill them. They are therefore useful for the treatment of acute infections such as boils, acne, vaginal thrush and other diseases of that nature, though care must be taken over the concentration, so as not to cause sensitisation (Pénoël & Franchomme, 1990, p.154).

Consentino et al. (1999)[44] researched the effects of three types of Thyme oil on various bacteria, and confirmed that it was the high phenolic content which caused these effects. Ultee et al. (1999)[45] researched the bactericidal mechanism of carvacrol in *Bacillus cereus* cultures, and found that carvacrol appears to interact with the cell membranes, causing them to increase in permeability to potassium and hydrogen ions. This increase in permeability causes a gradual impairment of bacterial cell processes, and leads to cell death.

- **Cholesterol lowering effects** - Case et al (1995)[46] found that dietary supplements of thymol and carvacrol (1mmol/kg) significantly reduced the serum cholesterol levels of cockerels. They suggested that it was due to the ability of the molecules to interact with the enzymes required for cholesterol manufacture by cells. This may have implications for human cholesterol levels, but it remains to be researched. It is probably unlikely that topically applied thymol and carvacrol would reach the blood stream in sufficient quantities to have significant effect.

---

[44] Cosentino S, Tuberoso CI, Pisano B, Satta M, Mascia V, Arzedi E, Palmas F. (1999) "In-vitro antimicrobial activity and chemical composition of Sardinian Thymus essential oils." *Letters in Applied Microbiology* Aug;29(2):130-5.

[45] Ultee A, Kets EP, Smid EJ. (1999) " Mechanisms of action of carvacrol on the food-borne pathogen Bacillus cereus." *Applied Environmental Microbiology.* Oct;65(10) pp. 4606-10.

[46] Case GL, He L, Mo H, Elson CE. (1995) "Induction of geranyl pyrophosphate pyrophosphatase activity by cholesterol-suppressive isoprenoids." *Lipids.* Apr;30(4) pp. 357-9.

Essential oils with high percentages of phenols

Ajowan *Trachyspermum copticum* L. Link	thymol 61%	para-cymene 15.5%	gamma-terpinene 11%
Bay (West Indian) *Pimenta racemosa* Mill. JS Moore	eugenol 56%	chavicol 21.6%	myrcene 13%
Cinnamon Leaf *Cinnamomum zeylanicum* Blume	eugenol 87%	benzyl benzoate 2.6%	beta-caryophyllene 1.8%
Clove Bud (Madagascar) *Eugenia caryophyllus* (Spreng.) Bullock	eugenol 76.6%	beta-caryophyllene 9.8%	eugenyl acetate 7.6%
Oregano (Greek) *Origanum vulgare* L. ssp. *viride* (Boiss.) Hayak (thymol chemotype)	thymol 85.6%	carvacrol 4.3%	gamma-terpinene 2.7%
Savory, summer (Italy) *Satureja hortensis* L	carvacrol 48%	gamma-terpinene 28%	para-cymene 7%
Thyme (Italy) *Thymus vulgaris*	thymol 27.4%	para-cymene 21.9%	gamma-terpinene 12%

ALDEHYDES

Structure - Aldehydes consist of an oxygen atom double bonded to a carbon atom (a carbonyl group C=O), at the end of a carbon chain. The fourth bond is always a hydrogen atom. They are derived from primary alcohols, by a process called oxidation.

Examples

trans-2-hexenal C_5H_9CHO Non-terpenoid Sweet Marjoram (0.3%) Clary Sage (0.05%)	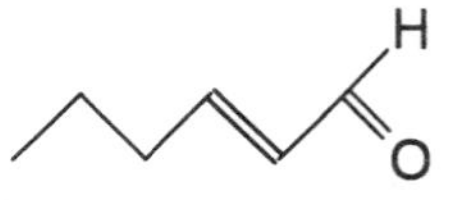
citronellal $C_9H_{17}CHO$ Monoterpenoid *Eucalyptus citriodora* (75%) Citronella (35%) *Litsea cubeba* May Chang (10%)	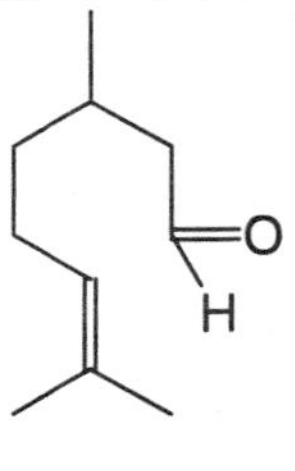

neral $C_9H_{15}CHO$ Monoterpenoid Melissa (35%) Lemongrass (35%) *Litsea cubeba* (30%) Ginger (3%) Lemon (1%)	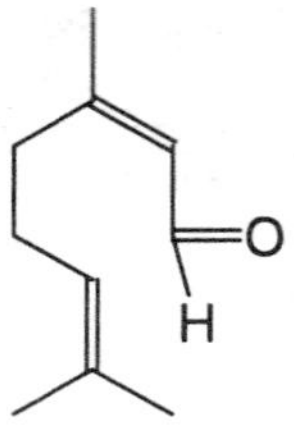
geranial[47] $C_9H_{15}CHO$ Monoterpenoid Lemongrass (50%) *Litsea cubeba* (25-40%) Melissa (20%) Ginger (10%) Lemon (2%)	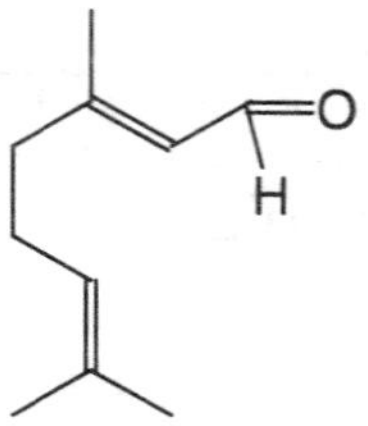

Naming - Aldehydes will either be called by their common name, followed by "**aldehyde**", for example **cumin aldehyde**; or they will end in "**-al**", for example cumin**al**. Aldehydes are derived from parent primary alcohols, eg. citronellal from citronellol.

Solubility - Aldehydes have a polar carbonyl group, making them slightly soluble in water. They are also soluble in ethanol and other oils.

Volatility - This depends on the structure of the constituent, but to generalise, it can be said that aldehydes are as volatile as alcohols, quickly being absorbed by the skin or by the nasal epithelium.

Reactivity - Aldehydes are very unstable and will readily oxidise to the acid form in the presence of oxygen and a little heat.

Toxic effects on the body - Aldehydes are often skin irritants, and sometimes will cause the eyes to weep, like onions do. Citral is irritating to the mucous membranes, as I found out once when I tried a gargle with one drop of Lemongrass oil to try and cure a sore throat. The pain experienced from the irritation of the citral was enough to make me forget about the original pain, but was not an experience I recommend. Tisserand & Balacs (1995, p.187) cite some cautions for use of citral-rich oils, in particular for glaucoma sufferers, as very low oral dosages (2-5

---

[47] Geranial and neral are so similar in chemical properties, that for a long while they were thought to be one molecule only, known as citral. These days citral often refers to a mixture of the two isomers.

micrograms) per day in monkeys caused an increase in ocular pressure. Another possibly malign effect are certain studies which showed citral causing overgrowth (hyperplasia) of prostatic[48] and vaginal[49] cells in rats after topical application at human-equivalent doses of about 10 ml.

Cinnamaldehyde, found in Cinnamon oils depresses rat liver glutathione levels, which may interfere with paracetamol (acetaminophen) metabolism, and cause liver toxicity if paracetamol is taken with Cinnamon oil (Tisserand & Balacs, 1995, p. 187).

Therapeutic effects of aldehydes

- **Calming properties** - Pénoël suggests (Pénoël & Franchomme, 1990, p.211) that aldehydes are calming to the central nervous system when inhaled, by "negativising" the nerve-endings in the olfactory system. A secondary calming effect on the alveolo-capillary interface of the lungs is the release of "calming" electrons into the serum of the blood. The understanding of Pénoël's theory of "negativant" and "positivant" molecules is explained in their book, but I have to confess it is a little esoteric for me. It may be interesting for readers to experiment with aldehyde-rich oils, in particular ones rich in citral, to see if they get a calming effect.

- **Anti-inflammatory** - again due to the release of electrons into the area of inflammation according to Pénoël (Pénoël & Franchomme, 1990, p.212).

- **Anti-infectious agents** - aldehydes are not so effective as the phenols, but do exhibit anti-infectious properties. The major exception to this rule is cinnamaldehyde, from cinnamon, which has anti-infectious properties on par with the phenols, and is also not calming.

  One use for aldehyde-containing oils could be in the management of opportunistic fungal infections in the last stages of AIDS, as Viollon

---

[48] Engelstein D, Shmueli J, Bruhis S, Servadio C, Abramovici A (1996) "Citral and testosterone interactions in inducing benign and atypical prostatic hyperplasia in rats." *Comparative Biochemistry & Physiology C - Pharmacology Toxicology and Endocrinology* 115(2) pp. 169-77.

[49] Geldof AA, Engel C, Rao BR. (1992) "Estrogenic action of commonly used fragrant agent citral induces prostatic hyperplasia." *Urology Research* 20(2):139-44.

and Chaumon (1994) suggest. The constituents that showed particular efficacy against *Cryptococcus neoformans* were phenols thymol and carvacrol, and also citral, and cinnamaldehyde[50].

- **Anti-melanoma activity** - Trans-2-hexenal, found in low concentrations in many of the leaf oils such as Oregano, Thyme, and Sweet Marjoram, is shown to inhibit glutathione-S-transferase enzymes in melanoma cells[51]. Glutathione is a molecule used by cells as an anti-oxidant, and if its production is inhibited, the cells are more likely to be killed by chemotherapy drugs.

Essential oils with high percentages of aldehydes

Cinnamon Bark *Cinnamomum zeylanicum* Blume	cinnamaldehyde 74%	eugenol 8.8%	cinnamyl acetate 5.1%
Citronella (Java) *Cymbopogon winterianus*	citronellal 36.8%	geraniol 21.4%	citronellol 15%
Lemon Balm/Melissa *Melissa officinalis* L.	geranial 45%	neral 35%	6-methyl-5-hepten-2-one 3%
May Chang *Litsea cubeba* (berry)	geranial 40%	neral 33.8%	limonene 8.3%

KETONES

Ketones also have an oxygen atom double bonded to a carbon atom, but always on a carbon atom which is bonded to two other carbon atoms. They are derived from secondary alcohols by oxidation. The oxidation of ketones usually takes longer than the oxidation of aldehydes, and mostly takes place in the plant.

---

[50] Viollon C, Chaumont JP (1994) "Antifungal properties of essential oils and their main components upon Cryptococcus neoformans. " *Mycopathologia* 128(3) pp. 151-3

[51] Iersel ML, Ploemen JP, Struik I, van Amersfoort C, Keyzer AE, Schefferlie JG, van Bladeren PJ. (1996)"Inhibition of glutathione S-transferase activity in human melanoma cells by alpha,beta-unsaturated carbonyl derivatives. Effects of acrolein, cinnamaldehyde, citral, crotonaldehyde, curcumin, ethacrynic acid, and trans-2-hexenal. *Chemical and Biolochemical Interactions* 102(2) pp. 117-32.

Examples

menthone $C_{10}H_{18}O$ Monocyclic Peppermint (30%) Geranium Bourbon (2%)	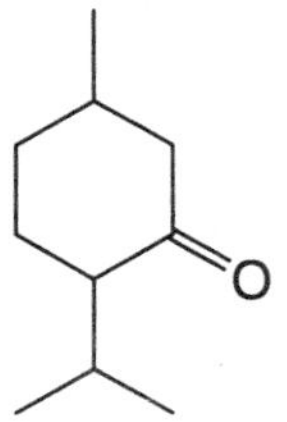
camphor $C_{10}H_{16}O$ Bicyclic Rosemary (15 - 30%) Sage (22%) Yarrow (12%) Spike Lavender (15%)	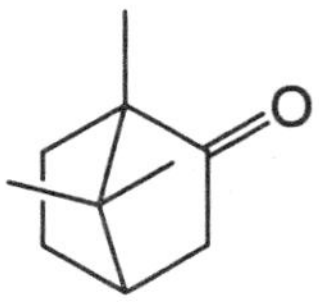
thujone $C_{10}H_{16}O$ Bicyclic Thuja/Cedarleaf (45%) Sage (20 - 40%) Wormwood (20%) Tansy *Tanacetum vulgare* (60%)	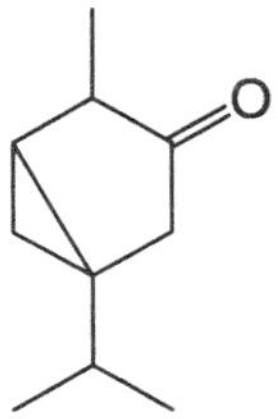

Naming - Ketones are either known by their common name, such as camphor, or will have the ending "**-one**". Ketones are derived from parent secondary alcohols, eg. menthone from menthol.

Solubility - The polar carbonyl group of ketones, makes them slightly soluble in water. They are also soluble in ethanol and other oils.

Volatility - Ketones have fairly high boiling points and tend to occur in crystalline form (e.g. camphor).

Reactivity - Ketones are relatively stable, and can pose problems in the body in that they are resistant to metabolism by the liver. The condition keto-acidosis suffered by people with cirrhosis of the liver is linked to an increase of ketone-type substances in the blood stream.

Toxic effects on the body - Some ketones are much more dangerous than any constituent we have come across so far. According to Tisserand & Balacs (1995, p.69) the ketones with potential convulsant or CNS damaging properties to be aware of are:

- **artemisia ketone** in Wormwood oil,
- **thujone** in Armoise, Dalmatian Sage (*Salvia officinalis*), Tansy, Thuja, Wormwood and Western Red Cedar (*Thuja plicata*) oils,
- **pulegone** in Buchu and Pennyroyal oils,
- **camphor** in Armoise, Ho leaf (camphor CT), Dalmatian Sage, and possibly Rosemary (camphor CT),
- **pinocamphone** and **iso-pinocamphone** in Hyssop oil

Rue oil also contains a ketone, 2-undecanone, but its CNS toxicity is unproven.

Therapeutic effects of ketones

- **Mucolytic agents** - According to Pénoël, (Pénoël & Franchomme, 1990, p. 194), ketones are very effective against mucosal secretions caused by disease (both respiratory and uro-genital). In particular, Bartholin gland cysts seem to respond to the use of oils high in ketones.

- **Wound-healing properties** - Pénoël recommends oils high in ketones for wounds, scars, burns and surgical wounds, suggesting that the ketones prevent cheloids and over-production of scar tissue ((Pénoël & Franchomme, 1990, p. 195).

- **Anti-haematomal properties** - Constituents containing two ketone groups (known as diketones) are supposed to have anti-haematomal properties, in particular the italidiones of the oil of *Helichrysum italicum ssp. serotinum* (Pénoël & Franchomme, 1990, p. 194). This means that they can help activate the white cells to clear up the dead red cells which cause the succession of rainbow colours in a bruise. However, I have not found any current references to this research.

Camphor (also known as 2-camphanone) has been shown to cause vasodilation of dog and rat portal veins by relaxation of the venous smooth muscle at dosages of 0.6 - 6 mg/kg of camphor in 0.1 ml of a carrier applied to the skin. The initial reason for the research was to determine whether camphor could alleviate hemorrhoidal bleeding and inflammation and the results suggest that perhaps camphor could be used for such a purpose, though with caution due to dermal sensitisation (Xie et al, 1992)[52].

- **Anti-viral properties** - Apparently papilloma virus, herpes viruses and other viruses which attack the nervous system such as shingles can be killed by ketones (Pénoël & Franchomme, 1990, p. 196). This remains to be confirmed in further scientific research.

Essential oils with high percentages of ketones

Artemisia alba (Belgium)	iso-pinocamphone 34.6%	camphor 21.1%	1,8-cineole 5.7%
Annual Wormwood (Yugoslavia) *Artemisia alba*	artemisia ketone 44.8%	1,8-cineole 9.6%	camphor 6.3%
Buchu *Barosma betulina* Bertl. (also *Agathosma betulina*)	(-)-menthone 35%	diosphenol 12%	(-)-pulegone 11%
Camphor (Japan) *Cinnamomum camphora* (Yellow or Brown camphor)	camphor 51.5%	safrole 13.4%	1,8-cineole 4.75%
Caraway (Netherlands) *Carum carvi* L.	(+)-carvone 50%	limonene 46%	cis-dihyrdocarvone 0.48%
Cedarleaf (Canada) *Thuja occidentalis*	alpha-thujone 56%	fenchone 15%	beta-thujone 14.7%
Hyssop (France)	iso-pinocamphone 32.6%	beta-pinene 22.9%	pinocamphone 12.2%
Spearmint (Greece) *Mentha spicata*	(-)-carvone 42.8%	dihydrocarvone 15.7%	1,8-cineole 5.8%
Pennyroyal *Mentha pulegium*	(+)-pulegone 63.5%	(+)-isomenthone 19.7%	(+)-neoisomenthol 5.7%
Rosemary (Spain) (camphor chemotype)	alpha-pinene 22%	camphor 17%	1,8-cineole 17%
Rue (Egypt) *Ruta graveolens*	2-undecanone 49.2%	2-nonanone 24.7%	2-nonyl acetate 6.2%

---

[52] Xie JM, Greenberg SS, Longenecker G. (1992) "Effects of 2-camphanone on canine portal vein blood flow and rat smooth muscle." *Gastroenterology*. 102(2) pp. 394-402.

Sage (Dalmatia) *Salvia officinalis*	alpha-thujone 37.1%	beta-thujone 14.2%	camphor 12.3%
Tansy (Belgium) *Tanacetum vulgare*	beta-thujone 50%	trans-chrysanthemyl acetate 20%	camphor 6.4%
Tumeric (Indonesia) *Curcuma longa*	turmerone 29.5%	ar-turmerone 24.7%	turmerol 20%
Mugwort (Germany) *Artemisia absinthum*	beta-thujone 46%	sabinyl acetate 25%	trans-sabinol 3.2%
Yarrow *Achillea millefolium*	camphor 17.7%	sabinene 12.3%	1,8-cineole 9.5%

ACIDS & ESTERS

Most acids from plants are so soluble in water that they do not occur in the essential oils, but in the waters of distillation. They are not terpenoid, but shorter chains, which accounts for this solubility. However, they are easily combined with alcohols to make esters. The acid hydroxyl group will readily leave if there are catalyst hydrogen ions present, and then the positively charged acid remnant will displace the hydrogen of the alcohol hydroxyl group, to produce the ester and a molecule of water.

Formation of the ester linalyl acetate from linalool and acetic acid.

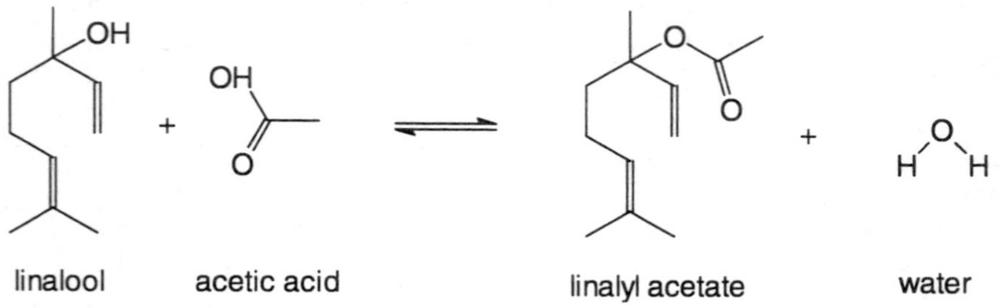

linalool   acetic acid   linalyl acetate   water

Naming - Acids which do occur in essential oils are usually carboxylic acids, and are terpenoid compounds. Thus you will get the ending "**-ic acid**", for example, salicy**lic acid**, which is found in Willow bark. Another familiar acid is acetic acid, which is found in vinegar, and often in the waters of distillation, but it is a two carbon chain molecule, and usually occurs in the conjugated ester form.

Esters are made from alcohols and acids, and are named after both parent molecules thus: "Linalool + acetic acid" become "Linalyl acetate". The alcohol drops the **-ol,** and gains a **-yl,** and the acid drops the **-ic,** and gains an **-ate.**

Examples

linalyl acetate $C_{10}C_{17}OCOCH_3$ Clary Sage (50%) Lavender (40%) Bergamot (25%)	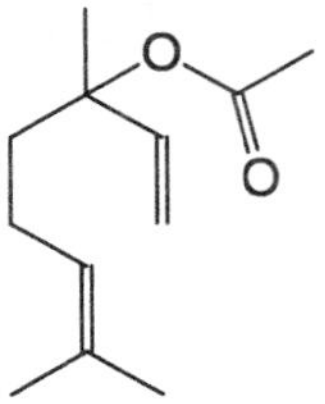
benzyl benzoate $C_7H_8OCOC_7H_8$ Narcissus absolute (20%) Jasmine absolute (16%) Ylang ylang (7%)	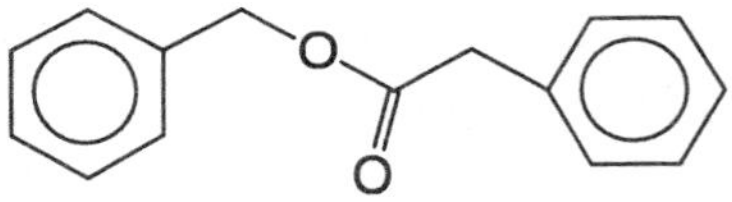

Solubility - Terpenoid acids are the most soluble in water of the essential oil constituents, due to their polar functional groups, but are still not very soluble (about 2 g/L). Esters are not very soluble, because the C-O-C part of their structure is neutralised by the opposing dipoles, and there are two non-polar carbon chains on the molecule, rather than just one.

Volatility - Again, this depends on structure and boiling point, but on the whole, acids and esters are not significantly more volatile than their equivalent alcohols. The confusing thing about esters is that they affect our nasal epithelia at very low concentrations, so we tend to think they are more volatile because they are more odorous. As we have seen, the other types of constituents also have quite strong smells too.

Reactivity - Esters are generally quite stable, particularly in essential oils where there is no water available for hydrolysis. However, it is still possible that the double bonds can oxidise as for the terpenes.

Toxic effects on the body - As the acids are easily metabolised by the body, and excreted in the urine, they do not generally pose a toxicity problem. As with all terpenoid molecules, esters and acids can cause skin sensitisation when used over long periods, and esters are held responsible for the drying effects on the skin, in oils such as Lavender, according to Dr. Pénoël (Pénoël & Franchomme, 1990, p.188). He also suggests that formates in high toxic doses may cause tachycardia, and acetates may cause epileptiform seizures, although there are no references quoted in the text.

Therapeutic effects of acids and esters

Acids are not given any special qualities by Dr. Pénoël, but they fall into the general category of being stimulants. Salicylic acid is known as an anti-inflammatory agent, but its side effects of gastric ulceration also put it in the stimulant category.

Esters

- **Anti-spasmodic and sedative properties** - these are the most important properties of esters, and according to Pénoël are thought to be due to the regulatory and re-equilibrating action of the esters on the sympathetic nervous system, and the neuro-endocrine system. He suggests that the intensity of the spasmolytic effect of an ester depends on the structure and number of carbons in the parent acid. That is, a benzoate ester with seven carbon atoms would be more spasmolytic than an acetate ester which has two carbon atoms. The level of action depends on the parent alcohol, as follows:

  Short chain alcohols give esters which are oriented towards the head, in particular the psyche. An example would be ethyl butyrate, where the "ethyl" part is derived from the alcohol ethanol.

  Monoterpenoid alcohols give esters which are oriented towards the regulation of the higher levels of metabolism. An example would be citronellyl acetate.

  Sesquiterpenoid alcohols give esters which are oriented towards the lower levels of metabolism and the genital sphere.

  Buchbauer et al. (1991)[53] experimented on mice with linalool, linalyl acetate and Lavender oil, and found that both linalool and linalyl

---

[53] Buchbauer G, Jirovetz L, Jager W, Dietrich H, Plank C. (1991) " Aromatherapy: evidence for sedative effects of the essential oil of lavender after inhalation." *Zeitschrift fur Naturforschung [C]*. Nov-Dec;46(11-12) pp. 1067-72

acetate exhibited a sedative effect on the mice, seeming to depress the motor cortex particularly.

Pénoël also mentions an anecdote where *Anthemis nobilis* (Roman Chamomile) oil used on the nape of the neck and carotid area of the neck had a similar effect to the pre-operative relaxing drugs, and could be used by people going for operations, with less chance of negative drug effects afterwards. Roman Chamomile oil contains large amounts of esters, particularly ones which have short chain alcohols as parent alcohols. The main constituent is isobutyl angelate (35%), which he also states is one of the "grand antispasmodics of the pharmacopoeia", and has the following structure:

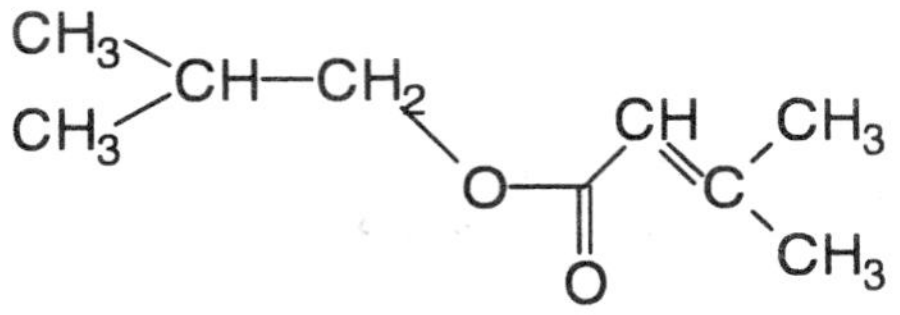

The other esters in Roman Chamomile oil are also isobutyl, isoamyl and propyl angelates and tiglates, thus all fitting into Pénoël's category of being derived from short chain alcohols and thus being cephalic and psychotropic (Pénoël & Franchomme, 1990, p.183).

- **Immuno-modulant effects** - there is some evidence that esters affect the levels of the globulin family, but the results are not conclusive as to whether the effect is suppression or stimulation (Pénoël & Franchomme, 1990, p.187).

Essential oils with high percentages of esters

Clary Sage (USA) *Salvia sclarea*	linalyl acetate 49%	linalool 24%	germacrene D 3%
Lavandin (Abrial quality, France) *Lavandula hybrida* Rev.	linalool 33.5%	linalyl acetate 27.1%	camphor 9.5%
Lavender (France) *Lavandula angustifolia* Mill.	linalyl acetate 40%	linalool 31.5%	(Z)-beta-ocimene 6.7%
Myrtle (Spain, wild growing) *Myrtus communis* L.	myrtenyl acetate 35.9%	1,8-cineole 29.9%	alpha-pinene 8.1%

Petitgrain bigarade *Citrus aurantium* L. ssp. *amara*	linalyl acetate 45.5%	linalool 24.1%	alpha-terpineol 5.2%
Roman Chamomile (Japan) *Anthemis nobilis*	isobutyl angelate 35.9%	2-methylbutyl angelate 15.3%	methallyl angelate 8.7%
Wintergreen (China) *Gaultheria procumbens*	methyl salicylate 90%	safrole 5%	linalool 2%

ETHERS

Structure

Most ethers occurring in the essential oils are phenolic ethers. They are derived from the hydroxyl group of the phenol (or alcohol), the hydrogen of which is replaced with a short carbon chain, either a methyl group (-CH_3), or a two carbon ethyl group (-CH_2-CH_3). The electronegativity of the oxygen atom causes opposite dipoles from the adjacent carbon atoms, so the net polarity of ethers is slight.

Examples of Ethers

methyl chavicol (chavicol methyl ether or methoxy allyl benzene) $C_9H_9OCH_3$ Basil (Comoro Islands) (85%) Tarragon (*Artemisia dranunculus*) (60%) Fennel (5%) Star Anise (*Illicium verum*) (5%)	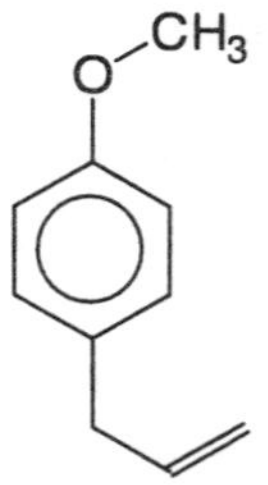
eugenol [54] $C_9H_9OHOCH_3$ Clove Bud (75%) Pimento (40%) Sweet Basil (5%) Rose (*Rosa damascena*) (1%)	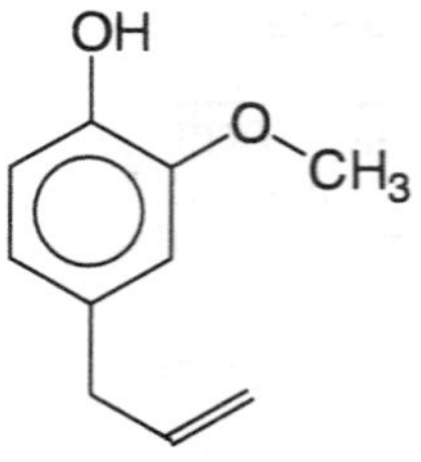

---

[54] Note that eugenol is both a phenol and a methyl ether. It will share properties of both, but the methyl ether group modulates the phenolic irritant effects.

trans-anethole
$C_9H_9OCH_3$
Aniseed *Pimpinella anisum* (93%)
Sweet Fennel (70%)
Star Anise (*Illicium verum*) (70%)

Naming

Ethers are phenols where the hydrogen atom of the -OH group has been exchanged for a carbon chain (usually a methyl group, $-CH_3$). They are often named after the phenol, plus the carbon chain group name, and with the word "ether" attached, e.g.: Chavicol methyl ether. "Chavicol" is the name of the phenol in this case (not an alcohol), "methyl" indicates a carbon chain of one, and "ether" indicates that they are linked by an oxygen group. In French and other languages, it is not uncommon to see the following abbreviation "Chavicol M.E.". The "M.E." stands for "methyl ether", as you have probably already guessed.

Another way of naming ethers is by using the carbon chain name (eg. methyl, ethyl, propyl, butyl etc.) and adding "-oxy" to indicate the presence of the oxygen group, eg: Meth**oxy** allyl benzene.

Sometimes ethers have an "**-ole**" ending, for example "estrag**ole**". This ending is also used for some oxides, and as we shall see later, oxides are somewhat similar to ethers in the arrangement of their oxygen atoms.

Solubility - Most ethers are soluble in ethanol, but as for all the long carbon chain molecules, they are negligbly soluble in water.

Volatility - The ethers tend to have boiling points below 200° C, and are thus fairly volatile compared to their cousins, the lactones and coumarins, most of which are solids at room temperature. Again this is dependent on the structure of the molecules, and the extent to which they experience inter-molecular attraction. Ethers tend to have very pungent smells, and can be detected at quite low percentages in an oil.

Reactivity - Ethers are not very reactive when exposed to heat and light, as both the methyl group and the benzene ring are resistant to oxidation.

The only changes which could occur are the changes to the C=C bond in the 3 carbon (propyl) "tail" of the molecule.

Toxic effects on the body

In large oral doses, ethers can have stupefying effects, leading to convulsions and even death. Tisserand & Balacs note that myristicin and elemicin, two ethers found in Nutmeg oil, are supposed to be psychotropic if ingested in sufficient quantities, but that they are fairly toxic at the doses which will produce the psychotropic effect (Tisserand & Balacs, 1995, p. 69-72, 194). Myristicin appears to inhibit the monoamine oxidase enzymes in the brain and increase the levels of brain serotonin, both of which would give rise to euphoric effects in humans. Trans-anethole had psychotropic effects in mice at an oral dosage level equivalent to 20 ml in humans.

It is unlikely that topically applied ether-rich oils would cause psychotropic effects, regardless of dosage, unless perhaps the person were already taking medication which altered the balance of serotonin and monoamine oxidase enzymes in the brain (for example certain anti-depressant medication).

Liver toxicity - In particular, the ether safrole found in yellow and brown *Cinnamomum camphora* oils, is noted for its liver carcinogenic effects in animals given daily oral doses over a long period of time (Tisserand & Balacs, 1995, p. 198). As with other constituents such as the ketone pulegone, it is actually the metabolites of the compound which are more carcinogenic than the original compound.

Therapeutic effects of ethers

These effects are only for the phenol methyl ethers, for example chavicol methyl ether and anethole. There are other types of ethers, for example, the spiroethers found in Clary Sage oil,which have not been researched.

- **Anti-spasmodic and analgesic** - They work on body organs below the diaphragm, in particular the colon, the genito-urinary tract and the muscles associated with them.

- **Anti-infectious** - the phenol methyl ethers have an all or nothing effect against micro-organisms, but most are effective against some type of micro-organism. Non-phenolic ethers (which do not occur in the family Labiatae) have anti-parasite action in the gut.

Essential oils with high percentages of ethers

Anise (Spain) *Pimpinella anisum*	(E)-anethole 96%	limonene 0.6%	anisaldehyde 0.56%
Basil (Comoro Islands)	methyl chavicol 85%	1,8-cineole 3.25%	para-cymene 2.7%
Sweet Fennel (Turkey) *Foeniculum vulgare* Mill. var. *dulce*	(E)-anethole 80%	limonene 6%	methyl chavicol 4.5%
Star Anise (China) *Illicium verum* Hook. F.	(E)-anethole 71.5%	foeniculin 14.5% (an azulene)	methyl chavicol 5%
Tarragon (USA) *Artemisia dranunculus* (French)	methyl chavicol 80%	beta-ocimene 14%	limonene 2.5%
Vanilla *Vanilla fragrans* Ames	vanillin 85% (ether)	4-hydroxybenzaldehyde 8.5%	4-hydroxybenzyl methyl ether 1%

OXIDES

Oxides are similar to ethers in that there is an oxygen atom in the carbon chain, but it is usually included in the formation of a ring, rather than being attached to a short carbon chain. The most ubiquitous is 1,8-cineole, or eucalyptol(e), as it is commonly known.

Examples of Oxides

1,8-cineole (eucalyptole) $C_{10}H_{18}O$ Bicyclic *Eucalyptus globulus* (70%) Cardamom (30%) Spike Lavender (15%) Sage (15%) Rosemary (15%)	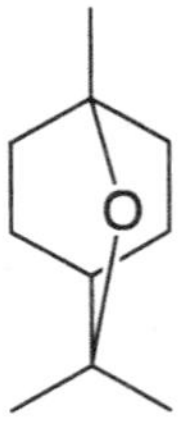
menthofuran $C_{10}H_{14}O$ Bicyclic Peppermint (4%) Cornmint (*Mentha arvensis*) (1%)	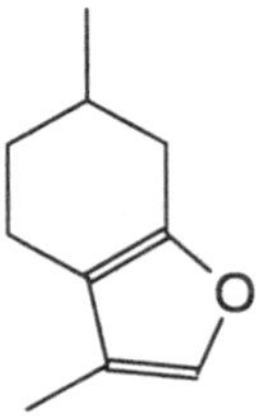

rose oxide $C_{10}H_{18}O$ Monocyclic Geranium (1%) Cistus (0.5%) Rose (*R. damascena*) (0.3%)	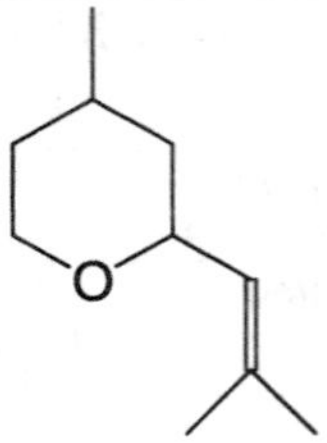

Structure - An oxide is formed where an oxygen atom is included in the structure to make a ring. As they are usually derived from alcohols, they often keep the alcohol name, and just add "oxide", eg: Linalol oxide. However, there are a few which go by their own special names, and the most common one is 1,8-cineole. The ending "-ole" is indicative of the oxygen atom in the structure. Other special cases are when a ring is made between two adjacent carbon atoms and the oxygen atom, or when the oxygen atom forms an unsaturated five-membered ring. The former are epoxides, and the latter are furans (see the diagram below). Examples of these are rare in the family Labiatae, but menthofuran is present in *Mentha piperita*.

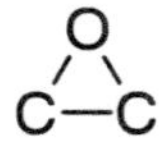

epoxide 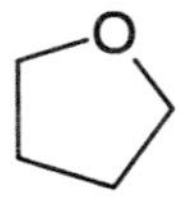   furan

Solubility - Similarly to ethers, the oxides are only slightly soluble in water, and dissolve fairly in other oils.

Volatility - Oxides also tend to have boiling points below 200° C, and are thus fairly volatile. Of the functional groups, they are possibly the strongest odorants, giving characteristic odours to the oils in which they occur, even at percentages as low as 0.3% (e.g. rose oxide in rose oil). Another oxide, lime oxide, is only present in steam-distilled Lime oil, and is the characteristic odour of that oil.

Reactivity - Oxides tend to decompose into alcohols under conditions of heat and light, and they can also form long chain polymers of terpenoid molecules, which end up forming a sticky residue. This is because oxides, in particular epoxides, exist in very strained conformations, and given a chance, will react to form a less strained structure. Epoxyresin, a preparation used in sealing wood, works on the principle of oxidising in the air to form a solid layer of interlinked molecules which started off as epoxides and have cross-reacted to form the solid seal.

Toxic effects on the body

Little research has been done on the oxides, with the exception of 1,8-cineole, which has similar neurotropic properties as the ethers, and has been known to cause ataxia, slurred speech, unconsciousness and convulsions[55,56]. Darben et al. (1998) cite a case where a six year old child experienced such symptoms after topical application of a home remedy for urticaria which contained eucalyptus oil[57].

Ascaridole is a terpenoid peroxide found in Wormseed (*Chenopodium ambrosiodes*) and Boldo (*Peumus boldus*) oil, which is toxic to humans in large doses, but in small doses has been used as an anti-parasitic agent for intestinal worms (Pénoël & Franchomme, 1990, p.190-191)

Menthofuran, found in trace amounts in Peppermint and other mint oils is also a metabolite of pulegone, a toxic ketone found in Pennyroyal oil. It destroys liver enzymes of the type cytochrome P450, which are responsible for detoxifying foreign substances, and as such can be very damaging to the liver[58].

Therapeutic effects of oxides

- **Expectorant effect** - Mucous glands and cilia of the respiratory tract are stimulated by oxides, in particular linalool oxide and 1,8-cineole, which results in an expectorant effect. Care must be taken with 1,8-cineole in asthmatics, as it can set off an attack. Oxides can also have a drying effect on the skin, if used often (Pénoël & Franchomme, 1990, p.189). Research by Ulmer & Schott (1991) supports the understanding that oxides have an expectorant effect. They conducted a double blind study of patients with chronic obstructive bronchitis, and found that an inhalation product called Gelomyrtol forte

---

[55] Day LM, Ozanne-Smith J, Parsons BJ, Dobbin M, Tibballs J. (1997) "Eucalyptus oil poisoning among young children: mechanisms of access and the potential for prevention" *Australian and New Zealand Journal of Public Health*. 21(3) pp. 297-302.

[56] Burkard PR, Burkhardt K, Haenggeli DA, Landis T (1999) "Plant-induced seizures: reappearance of an old problem." *Journal of Neurology* 246(8) pp. 667-670.

[57] Darben T, Cominos B, Lee CT (1998) " Topical eucalyptus oil poisoning." *Australasian Journal of Dermatology* 39(4) pp. 265-7

[58] Tisserand R & Balacs T (1995) "*Essential Oil Safety*" Churchill Livingstone, England p. 193.

containing 1,8-cineole decreased coughing, increased ease of expectoration, and decreased shortness of breath. in the experimental group as compared to the control, and also improved the colour of the sputum over the 14 day treatment period[59].

Juergens et al. (1998) suggest that 1,8-cineole exhibits anti-inflammatory effects in bronchial asthma by inhibiting the leukotriene B4 and prostaglandin E2 pathways in blood monocytes of humans with bronchial asthma. The effect was noted in monocytes of healthy volunteers as well as the asthma sufferers[60].

Essential oils with high percentages of oxides

Eucalyptus globulus (Spain)	1,8-cineole 66.1%	alpha-pinene 14.7%	limonene 3%
Cajeput *Melaleuca leucadendron* L.	1,8-cineole 41.1%	alpha-terpineol 8.7%	para-cymene 6.8%
Cardamom (Reunion) *Cardamomum elettaria* L. Maton var. *alpha-minor*	1,8-cineole 48.4%	alpha-terpinyl acetate 24%	limonene 6%
Niaouli (Madagascar) *Melaleuca quinquinervia* Cav.	1,8-cineole 41.8%	viridiflorol 18.1%	limonene 5.5%
Myrtle (Spain, wild growing) *Myrtus communis* L.	myrtenyl acetate 35.9%	1,8-cineole 29.9%	alpha-pinene 8.1%
Rosemary (Tunisia) (probably cineole chemotype)	1,8-cineole 51.3%	camphor 10.6%	alpha-pinene 10%
Rosemary (Spain) (? camphor chemotype)	alpha-pinene 22%	camphor 17%	1,8-cineole 17%
Spike Lavender (Spain) *Lavandula latifolia* Medicus	1,8-cineole 36.3%	linalool 30.3%	camphor 8%
Angelica (root) *Angelica archangelica*	alpha-pinene 25%	1,8-cineole 14.5%	alpha-phellandrene 13.5%

[59] Ulmer WT, Schott D (1991) "Chronic obstructive bronchitis. Effect of Gelomyrtol forte in a placebo-controlled double-blind study" *Fortschritte der Medizin* September 20, 109(27) pp. 547-550.

[60] Juergens UR, Stober M, Schmidt-Schilling L, Kleuver T, Vetter H (1998) "Anti-inflammatory effects of eucalyptol (1,8-cineole) in bronchial asthma: inhibition of arachidonic acid metabolism in human blood monocytes ex vivo" *European Journal of Medical Research* 3(9) pp.407-412.

LACTONES

Structure - Lactones resemble ketones, esters and ethers, and are probably derived from a condensation reaction of an alcohol and an aldehyde on two different carbon atoms. They can be either monoterpenoid or sesquiterpenoid, and always have a C=O next to an oxygen atom that is part of a closed ring.

Examples

helenaline (Tricyclic) $C_{15}H_{18}O_4$ *Arnica montana* (in alcoholic extract)	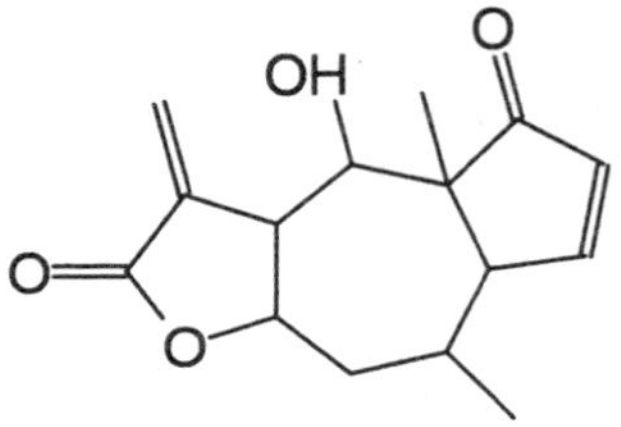
nepetalactone (Bicyclic) $C_{10}H_{14}O_2$ *Nepeta cataria* Catnip (80%)	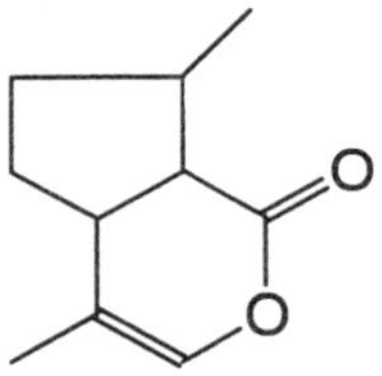
alantolactone (Tricyclic) $C_{15}H_{20}O_2$ Sweet Inule or Elecampane *Inula graveolens* (20%)	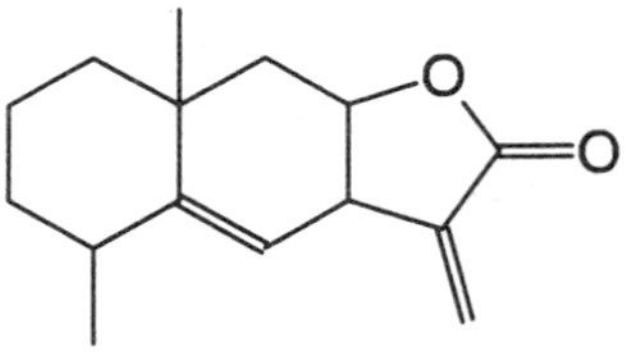

Naming - Lactones are usually known by their common names, as the chemical ones are too lengthy. They tend to end in "**-lactone**", but can also end in "**-in**", or "**-ine**".

Solubility - Lactones are not particularly soluble in water, as most of the other essential oil constituents.

Volatility - Monoterpenoid lactones are more volatile than sesquiterpenoid ones, as might be expected. Sesquiterpenoid lactones are not very volatile, and their often earthy and pungent odours linger for a long time.

Reactivity - Lactones have similar reactive properties to ethers and to ketones, and do not easily oxidise. They are also not metabolised particularly easily, and some have adverse effects on the metabolic process (see below).

Toxic effects on the body

Lactones tend to have the same neurotoxic effects as ketones according to Dr Pénoël (Pénoël & Franchomme, 1990, p.203), and also can cause skin allergies, sensitisation or blistering. In particular, plants from the Compositae (Asteraceae) family have sesquiterpenoid lactones which are responsible for a large bulk of the allergic contact dermatitis some people have to daisy-type flowers, particularly chrysanthemums[61]. It is interesting to note that alantolactone (cited as a mucolytic by Dr. Pénoël) is also among those listed as a sensitising lactone. Fortunately, lactones only occur in small amounts in a few essential oils. Some people have reported allergies to chamomile oil (both Roman and German)[62,63] that suggest a lactone-type reaction, and there are traces of lactones in both oils.

Another perhaps more serious effect of some sesquiterpene lactones, including helenalin found in *Arnica montana,* is the reduction in liver enzyme activity in rats[64]. In particular, the glutathione-S-transferase and NADPH-cytochrome P450 reductase enzymes were affected. These two enzymes are responsible for keeping glutathione and cytochrome P450 levels high enough to protect the liver from toxic substances, including free radicals and drugs like paracetamol, which are liver toxic[65]. It is unlikely that at the levels of oils used in aromatherapy that there would be any effects noticed, but it is interesting to note the possibility.

---

[61] Gordon LA (1999) "Compositae dermatitis" *Australasian Journal of Dermatology* 40(3) pp. 123-128.

[62] Van Ketel WG (1982) "Allergy to Matricaria chamomilla" *Contact Dermatitis* 8(2) pp. 143

[63] McGeorge BC, Steele MC (1991) "Allergic contact dermatitis of the nipple from Roman chamomile ointment" *Contact Dermatitis* 24(2) pp. 139-140.

[64] Joydnis-Liebert J, Murias M, Bloszyk E (2000) "Effect of sesquiterpene lactones on antioxidant enzymes and some drug-metabolizing enzymes in rat liver and kidney" *Planta Medica* 66(3) pp. 199-205.

[65] Tisserand R & Balacs T (1995) "*Essential Oil Safety*" Churchill Livingstone p.38, 42-3.

Therapeutic effects of lactones

- **Mucolytic and expectorant** - In particular, the lactones alantolactone and isoalantolactone found in *Inula graveolens* (Sweet Inule) and *I. helenium* (Elecampane) are very effective in treating chronic catarrah and bronchial congestion, due to the heightening of the ketone function by the nearby oxide group (Pénoël & Franchomme, 1990, p.202). Dr. Pénoël records a case study where the inhalation of the oil caused a "healing crisis", but within two days completely cleared a chronic bronchial and sinus infection.

  Mazor et al. (2000) studied the effects of isohelenin (also known as isoalantolactone) and parthenolide on the inflammatory process in human respiratory epithelium. They found that the lactones inhibited the expression of the gene for interleukin 8, a molecule which promotes the inflammatory process in human respiratory epithelium, and suggest that this may be one mechanism whereby the lactones exert their anti-inflammatory effect[66]. It may be that the results Dr. Pénoël noted with the use of *Inula graveolens* oil were more due to the anti-inflammatory effect of the isoatlantolactone than to its mucolytic or expectorant effects.

- **Anthelmintic** - (anti-worm) various lactones are effective for different types of worms, but none occur in the family Labiatae.

- **Anti-inflammatory** - In spite of some lactones being allergens for some people, other research is showing that sesquiterpenoid lactones have powerful anti-inflammatory properties. Helenalin, another sesquiterpenoid lactone this time from *Arnica montana* extracts, not the oil (although there may well be a market for a CO_2 extract of Arnica), showed a similar anti-inflammatory effect to that of isohelenin (isoatlantolactone - don't get the names confused!). Again, it is the inhibition of the transcription factor NF-kappaB which mediates the anti-inflammatory effects, shown here in T-cells, B-cells and epithelial cells[67]. This is a different mechanism than that used by aspirin and other nonsteroidal anti-inflammatory drugs.

---

[66] Mazor RL, Menendez, IY, Ryan MA, Fiedler MA, Wong HR (2000) "Sesquiterpene lactones are potent inhibitors of interleukin 8 gene expression in cultured human respiratory epithelium" *Cytokine* 12(3) pp. 239-245.

[67] Lyss G, Schmidt TJ, Merfort I, Pahl HL (1997) "Helenalin, an anti-inflammatory sesquiterpene lactone from Arnica, selectively inhibits transcription factor NF-kappaB." *Biological Chemistry* 378(9) pp. 951-961.

- **Possible analgesic effects** - Aydin et al. (1998) examined nepetalactone, a monoterpenoid lactone found in *Nepeta caesarea* (Catnip sp.) essential oil. They suggest that oral dosages provide not only a marked sedation in rats, but also mediated the rats response to mechanical pain[68]. This is not the same as having pain in the first place, and then using an analgesic to stop the pain, but nevertheless is of some interest. They also suggest that nepetalactone has a specificity for particular opioid receptors, which would put it in the same category as analgesics such as codeine.

Essential oils with high percentages of lactones

Catmint *Nepeta cataria*	nepetalactone 1 80.5%	nepetalactone 2 10%	beta-caryophyllene 4.4%

No essential oils used in aromatherapy have lactones as their major constituents. This is probably just as well, due to their potential skin sensitising properties.

COUMARINS

Structure - Coumarins are a type of lactone, and both share the functional group of a C=O joined onto a carbon which is bonded to another oxygen atom. Coumarins also have the distinctive feature of an adjacent benzene ring, which in turn can have several different functional groups attached.

Examples

coumarin (Bicyclic) $C_9H_6O_2$ (not quite terpenoid) Tonka Bean (50%) Cassia (12%) *Narcissus poeticus* absolute (6%)	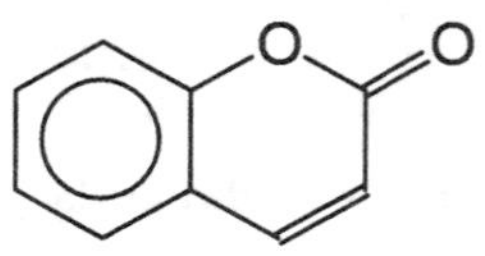

---

[68] Aydin S, Beis R, Ozturk Y, Baser KH (1998) "Nepetalactone: a new opioid analgesic from Nepeta caesarea Boiss." *Journal of Pharmacy and Pharmacology* 50(7) pp. 813-817.

herniarin (7-methoxy-coumarin) $C_9H_6(OCH_3)O_2$ Bicyclic Tarragon (*Artemisia dranunculus*) 0.2% Lavender (trace)	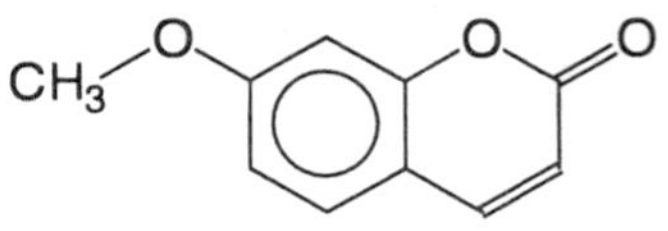
umbelliferone (7-hydroxycoumarin) $C_9H_6(OH)O_2$ Bicyclic Dill *Anethum graveolens* (2%)	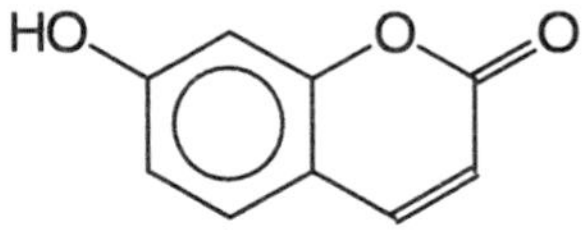

Naming - Coumarins also usually known by their common names, as the chemical ones are too lengthy. They tend to end with "**-in**" e.g. herniarin, but can also end in "**-one**", e.g. umbelliferone, otherwise known as 7-hydroxy-coumarin. Some researchers refer to coumarins as "benzopyrones", where the "-pyrone" ending indicates the presence of a six membered ring with an oxygen atom as part of the ring.

Solubility - Coumarins are less soluble in ethanol than most other constituents, and are negligbly soluble in water.

Volatility - Tend to be solids at room temperature. Again this is dependent on the structure of the molecules, and the extent to which they experience inter-molecular attraction. They would be more prevalent in solvent extracted oils as they are not volatile enough to be extracted in high quantities by steam distillation.

Reactivity - Coumarins do not readily oxidise. They are also not metabolised particularly easily, due to the benzene ring and the resistance of the lactone part to oxidation.

Toxic effects on the body

A chemical (not found in essential oils) known as dicoumarol is a potent anticoagulant. It is generated in the stomachs of herbivores that have eaten too much Sweet Clover which contains coumarin, and causes abnormal internal bleeding[69]. Other coumarins may also interfere with blood-clotting, although this has not been proven conclusively. It is as well to avoid oils with coumarins in if a patient is undergoing warfarin

---

[69] Lehninger A, (1982) "Principles of Biochemistry" Worth Publishers, NY, p.276.

treatment for a thrombosis. Note the coumarin group present in the structure of warfarin below:

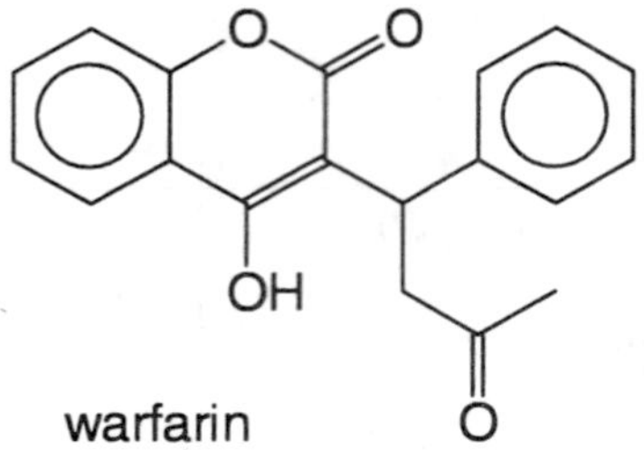

warfarin

The well-known UV sensitising effect of bergamot oil on the skin is due to the presence of a furocoumarin, bergapten(e), which links up with the DNA of cells responsible for manufacture of melanin. It does this because its chemical structure can align with the pyrimidine bases in the DNA molecules.

Once linked, the structure of bergaptene (benzene ring next to an unsaturated furan ring) allows for the production of a free radical oxygen in the presence of UV light which can inactivate important enzymes in the cell, and cause peroxidation of unsaturated lipids[70]. This means that the UV rays can cause much more damage because the cell's normal protection mechanisms (enzymes) are out of action. It is unclear exactly why the bergapten should cause a darkening of the affected skin, though it is no doubt the body's response to the deep cellular damage which occurs.

Therapeutic effects of coumarins

- **Sedative effects** - these are due to the desensitising of the nervous reflex systems. Coumarins can have hypnotic effects in large doses, and are known as sleep-inducers according to Dr. Pénoël (Pénoël & Franchomme, 1990, p. 205).

- **Anti-lymphoedemic effects** - Casley-Smith (1999) has reviewed fifty trials of different coumarins used in the treatment of lymphoedema, some of which were oral dosages, others topical. Oral coumarin treatment significantly reduced oedema by up to 55% as compared to control groups, and some trials went over several months, the effects being greater for the greater length of time.

[70] Grossweiner LI (1984) "Mechanisms of photosensitization by furocoumarins" *National Cancer Institute Monographs* December v. 66 pp.47-54.

Topically applied coumarins also reduced oedema, but not as effectively. The only side-effect noted was idiosyncratic hepatitis (3 in 1000 likelihood)[71].

Burgos et al. (1999) reported use of coumarin in 90 mg/day oral dosages with women suffering from lymphoedema as a result of surgical removal of breast cancer. During 12 months of coumarin therapy, the oedema significantly improved, with the volume of the arms decreasing, and the heaviness, hardness and discomfort in the nerves reducing as well[72].

Essential oils with high percentages of coumarins

Tonka Bean *Dipteryx odorata*	coumarin about 50%	?	?

Tonka Bean oil is not strictly an oil, but is a crystalline residue found in the pods of the Tonka Bean tree (a Brazilian rainforest tree). Mrs Grieve suggests that Tonka Bean extract is a cardiac, tonic and narcotic when used herbally, and that large doses can paralyse the heart (Grieve, 1976[73]). This is no doubt contributed to by the high percentage of coumarin in it, and she does not list any other constituents. Coumarin was used in the perfumery industry until it was identified as a possible carcinogen in the 1950s.

No essential oils used in aromatherapy have high percentages of coumarins. The cold pressed citrus oils contain furocoumarins (e.g. bergapten), up to about 3% in Bergamot oil (the others only have traces). Rue oil also contains about 7% bergapten. Simple hydroxycoumarins are found in small amounts in the following oils: Dill, Lavender, Tarragon, German Chamomile, Melissa (Pénoël & Franchomme, 1990, p.208). Dr. Pénoël also lists some more complex coumarins on p. 208, which I have not been able to find structures for so far: osthole,

[71] Casley-Smith JR (1999) "Benzo-pyrones in the treatment of lymphoedema" *International Angiology* 18(1) pp. 31-41.
[72] Burogs A, Alcaide A, Alcoba C, Azcona JM, Garrido J, Lorente C, Moreno E, Murillo E, Olsina-Pavia J, Olsina-Kissler J, Samaniego E, Serra M (1999) "Comparative study of the clinical efficacy of two different coumarin dosages in the management of arm lymphedema after treatment for breast cancer" *Lymphology* 32(1) pp. 3-10.
[73] Grieve M (1976) *A Modern Herbal*, Penguin Books, England, p.819

phellopterine, angelicine and umbelliprenine in Angelica oil; auraptene in Grapefruit oil; visnadine in Visnage oil (*Ammi visnaga*).

CHAPTER FIVE

BODY SYSTEMS & ESSENTIAL OILS

When using essential oils on the body, whether for medicinal or aesthetic purposes, it is important to know how they are going to:

- Get into the body
- What will happen to the constituents, and to the tissues and organs of the body once the constituents are inside
- How long they will stay in, and the mechanisms of excretion.

These type of questions are answered best by the study of biochemistry and pharmacology, and even molecular biology, all of which are fascinating fields of study in their own right. However, for the purposes of this book, I examine some of these areas from a lay person's point of view, explaining the interactions of the oils and the body systems as simply as possible.

We start with a description of some of the chemical reactions and essential oil reactions in cells. We then examine how the skin reacts to essential oils, and the extent to which it allows the passage of essential oils into the body and briefly examine the respiratory system as an alternative method of uptake for systemic absorption of the oils.

Then we will look at how the body deals with the essential oils at the level of the liver and kidneys, and summarise what seem to be the main types of reactions that essential oils provoke or inhibit in the body. This is followed by a brief examination of the way essential oils affect the olfactory system, and some suggestions as to the action of oils on mood and the immune system.

I acknowledge that there is a host of information about the esoteric and subtle use of the essential oils, but it is beyond the scope of this book. Some people are fond of the chart invented by Dr. Pénoël which is a quadrant square with two axes, indicating polarity or non-polarity and electronegative or electropositive character of the constituents. A leap is then made to assign the Chinese Yin-Yang system and words like hot, cold, wet and dry to the four quadrants, which are actually chemically defined, not esoterically defined. While there may be some interesting associations to be made with this approach, I encourage people to bear in mind that it is a model which may be helpful, but need not necessarily be true in all respects.

Cells

In every tissue and organ in the body there are cells, which are the basic units of life within the body. Every cell has particular functions and features, but there are some functions and features which are common to all cells.

At the cellular level, the things that you can observe happening are chemical reactions. It is at this level that the essential oil molecules will react chemically with the normal (or diseased) functioning of cells and cause the changes. Below is an overview of the types of chemical reactions that every cell does, and some suggestions as to how essential oil molecules may affect them.

- **Respiration** - Ordinary usage of the word respiration refers to the actions needed for breathing. At a cellular level, respiration is where the oxygen which has been absorbed during breathing is used in chemical reactions with glucose and other sugars to provide energy for the other chemical reactions in the cell. This energy is stored in the form of adenosine tri-phosphate (ATP) molecules which in turn easily lose a phosphate group to become adenosine di-phosphate (ADP) releasing chemical energy. Most enzymatic reactions require at least one molecule of ATP to be transformed into ADP to make the reaction go ahead.

  The reason why it is so important that the brain is not deprived of oxygen is because so many chemical reactions are going on all the time in the brain, that a lack of oxygen will quickly interrupt the normal functioning of the brain cells, often leading to irreversible chemical damage and death (at least for the cell).

  The waste products of respiration include carbon dioxide (CO_2), which in turn is used by plants to make glucose (as we saw in Chapter Two).

- **DNA transcription and manufacture of proteins** - Deoxyribose nucleic acid (DNA) is a huge molecule made up of sequences of smaller molecules known as nucleotides (adenine, guanine, thymine, cytosine, and uracil found in RNA), which are joined together by a ladder structure of ribose sugar phosphate molecules). The pattern of the nucleotides stores the genetic information in the DNA, but until the DNA molecule undergoes a process called transcription, the genetic information is not released. The ribonucleic acid (RNA) molecules form themselves into special 3-D structures known as ribosomes for the purpose of "transcribing" the genetic information off the DNA and making proteins from amino acids using the DNA pattern as a template.

One problem which cells face is how to repair damage to DNA, and fortunately there are some specific molecules whose job it is to do that. If the DNA is not repaired, then the genetic information is altered, which leads to transcription of a protein which may or may not function normally. Cancer is linked to mutations in the DNA which lead to the production of the wrong type of proteins, which then promote abnormal cell division and cell growth. As we saw in Chapter Four, the group of essential oil molecules known as psoralens (e.g. bergapten) can interfere with the DNA molecules in skin cells, causing damage, though not necessarily cancer.

- **Enzyme catalysed reactions** - Protein molecules fulfill many different functions in the cell, and are the most common type of molecule in cells. A large number of proteins function as enzymes. An enzyme is a molecule which is necessary for the chemical reaction of two or more other molecules, but it is not changed itself in the reaction. If a person is deficient in a certain enzyme, then they will often be unable to make certain other molecules required for their cells' functioning, or a toxic by-product is made instead of a useful molecule.

An example is the genetic disease known as phenylketonuria (PKU). Phenylalanine is an amino acid found in certain foods which is normally converted into another amino acid called tyrosine by an enzyme called phenylalanine hydroxylase (N.B. the "-ase" ending indicates an enzyme). If this enzyme is not present, the cell will use a different enzyme to try and cope with the phenylalanine. This secondary pathway results in the formation of a molecule called phenyl pyruvate, which has a ketone group attached to it (hence "phenyl**keton**uria"). The body cannot do anything with phenyl pyruvate except store it in the cells, or try and excrete some of it in the urine (hence "phenylketon**uria**"). Excess phenyl pyruvate in the blood during a child's early years prevents normal brain development, and results in severe mental retardation[74].

Essential oils do appear to interfere with certain enzyme functions and production, in particular during the inflammatory response, as we shall see later. It remains to be seen how specific this interaction is to particular enzymes.

- **Making structural molecules** - Other proteins form part of the structure of the cell or the substance produced by the cell e.g. collagen in skin, dentine in the teeth, or keratin in hair and nails. These products are usually made by epithelial cells.

- **Responding to information from outside the cell** - Some proteins perform special "gate-keeper" functions in the membranes of cells. They allow cells to communicate with one another, and to take in essential nutrients (such as water, salts like potassium and sodium, glucose and amino acids) and excrete waste products (such as urea and carbon dioxide).

Olfactory proteins
Our sense of smell depends on the family of G-proteins which are studded in the membranes of our olfactory neurons. When a "smelly" molecule comes in contact with an olfactory G-protein, the protein will register what shape molecule it is (or perhaps the molecule's bond energy vibrational pattern), and if it fits, the G-protein will change shape on the inside of cell. This change of shape triggers certain enzymes to catalyse other chemical reactions which generate an electrical impulse. This impulse is carried along the membrane of the neuron, transmitted to other neurons in the limbic system and the brain, and is finally interpreted as a smell by that part of us which is self-aware.

Adrenaline receptors
Other "gate-keeper" proteins are hormone receptors, which tell the cell to respond in a certain way if the correct hormone binds to the receptor. One very important hormone which all smooth and skeletal muscle cells in the body respond to is adrenaline (or epinephrine). If the adrenaline receptor proteins are blocked by another molecule, then

---

[74] Lehninger A (1982) "Principles of Biochemistry" Worth Publishers, NY,Chapter 19, p. 541

the adrenaline's "stimulate" message is not received. Synthetic drugs which block adrenaline receptors in heart muscle are known as "beta-blockers" because the heart muscle has adrenaline receptors knowns as "beta-receptors", which are blocked by certain chemicals, and not by others (Tortora & Grabowski, 1993[75]). Some essential oils may be found to have similar "blocking" or antagonistic properties.

Oestrogen receptors
One property I often see mentioned for essential oils is that they are "emmenagogic" or "oestrogenic". This would imply that they interact in some way with the reproductive hormonal cycle, either mimicking the action of oestrogen, or blocking production of progesterone. Below are the structures of beta-estradiol, the main oestrogen molecule, and progesterone, which also controls the reproductive cycle. Further below are the structures for anethole and diethylstilbesterol. Diethyl stilbesterol was once used as a mimic for beta-estradiol, and Dr. Pénoël suggests that anethole from Aniseed oil could also work like diethyl stilbesterol (Pénoël & Franchomme, 1990, p.175).

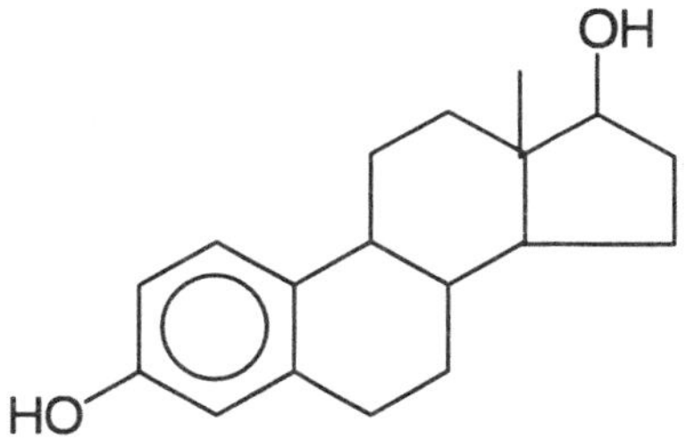

beta-estradiol

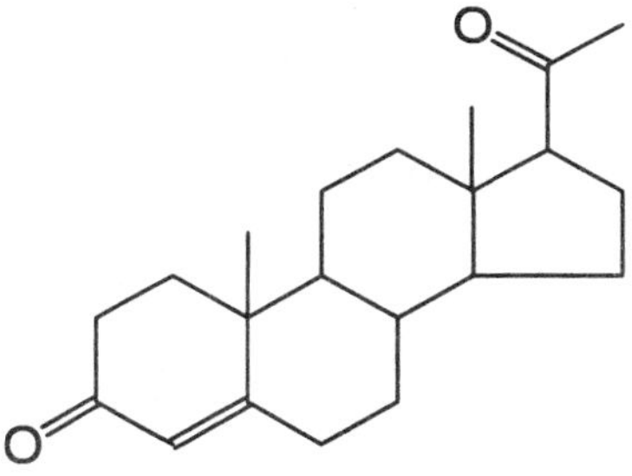

progesterone

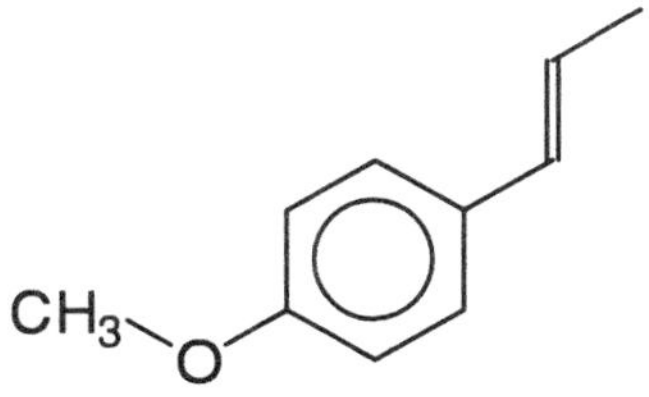

anethole

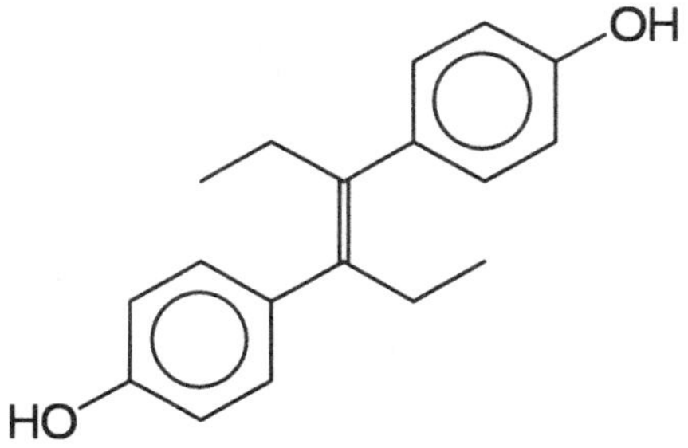

diethylstilbesterol

---

[75] Tortora GJ & Grabowski SR (1993) *"Principles of Anatomy and Physiology"* 7th Edition, HarperCollins College Publishers, NY, p.511

I don't know if there is enough evidence just from the structural similarities for saying that anethole will have oestrogenic action in the body. However, one of my students was astounded when I mentioned this fact, and shared with the class that she had been wearing aniseed essential oil, drinking fennel tea for several days, eating aniseed sweets, licorice, roast fennel vegetable and basil pesto because she loves the aniseed flavour, and had been surprised when her menstrual period came a week earlier than she expected.

- **Maintenance of phospholipid membrane** - every cell has a phospholipid membrane which encloses the cell, keeping its contents distinct from the watery fluid between the cells. This membrane can be damaged with reactive molecules known as "free radicals" and other types of molecules such as detergents which dissolve the lipids.

Free radicals are molecules which have one less atom bonded to them than is required for molecular stability, and will "attack" other molecules which have C=C double bonds to try and get access to another electron to share and become stable. The phospholipids in membranes are susceptible to attack by free radicals (which are found in epoxy-resin fumes, car exhaust, fried foods and oils which are rancid or "off").

Molecules such as alpha-tocopherol (also known as Vitamin E) are known as anti-oxidants because they interrupt the oxidation chain reaction started by the radicals. Notice the similarity in structure of Vitamin E to the phenols or phenyl methyl ethers.

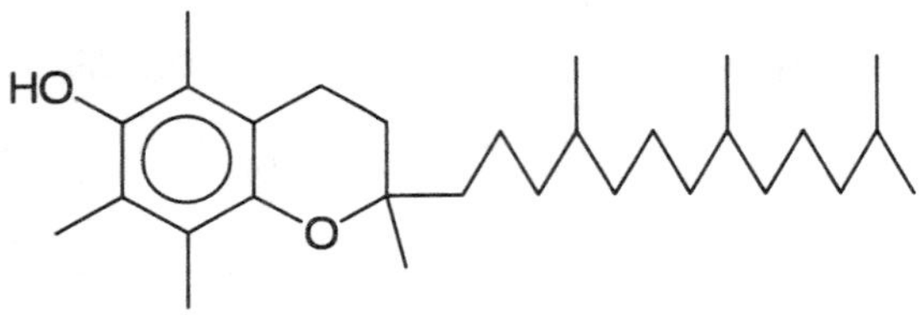

alpha-tocopherol (Vitamin E)

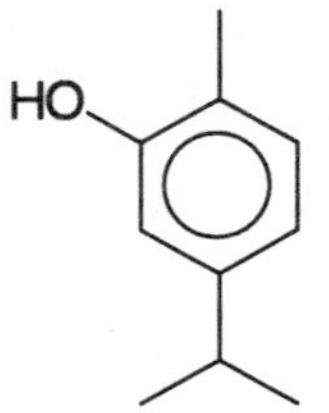

carvacrol 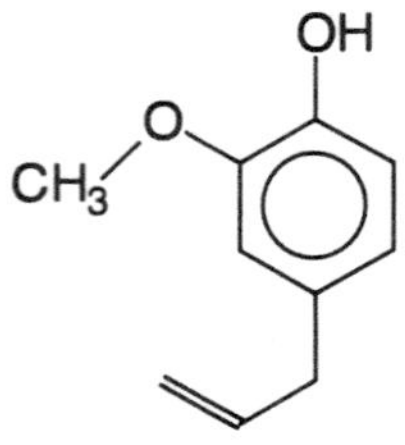   eugenol

It has been suggested that certain essential molecules such as carvacrol, thymol and eugenol also exhibit anti-oxidant properties (Aeschbach et al., 1994[76])

SKIN

An understanding of the structure and function of the human skin will give us an understanding of how the essential oils penetrate and are excreted through it. The skin is basically an organ which "keeps the outside out, and the inside in". Our bodies are made up of many different types of cells. In order to stay alive, cells require a certain type of environment:

- They must be in a fluid medium at all times, which contains water, salts, nutrients and chemical messengers from other cells to keep everything working together.
- In mammals, they must be at a constant temperature, to allow the necessary inter-molecular reactions to proceed (remember the necessity for extra energy for the formation of new bonds between molecules)
- They must be kept separate from toxic chemicals, infections, parasites or energy like UV light which could damage or destroy them.

The skin participates in all of the above functions with remarkable constancy, due to its complex layers of cells and extracellular fibres. Without skin, we would dry up, freeze and be exposed directly to the toxic chemicals and germs in our environment. The skin is also a major sensory organ, and a protection layer against harmful free energy in the form of ultra-violet light.

The skin has three major layers, the epidermis, dermis and subcutaneous layer (which is strictly speaking not skin anymore, but is included here as it contains the adipose tissue). Each layer of the skin has different characteristics which affect the movement of essential oils through them.

Epidermis

To translate the word, **epi-** means outermost, and **-dermis** means skin, so this is the outermost layer, and is responsible for preventing water loss. It

---

[76] Aeschbach R, Loliger J, Scott BC, Murcia A, Butler J, Halliwell B, Aruoma OI (19940 "Antioxidant actions of thymol, carvacrol, 6-gingerol, zingerone and hydroxytyrosol." *Food & Chemical Toxicology*.32(1) pp.31-6.

is about 0.2 mm thick, and is largely made up of dead cells which are filled with a substance called keratin (found in the outermost part of the epidermis known as the stratum corneum). Keratin is relatively inpermeable to water, and thus prevents the movement of water in or out through the skin. These dead cells are continually being rubbed off, and replaced by the reproductive layer in the epidermis.

The natural sebaceous oils allow the epidermis to expand when immersed in water, but they can be dissolved and removed from the skin by solvents such as ethanol and detergents. When this happens, or the sebum is depleted by a lack of intake of essential fatty acids in the diet, the epidermis loses its own moisture, and can split and crack. Wrinkles are actually formed in the dermal layer, where the molecules of collagen and elastin are found, but lack of sebaceous oils contributes to the drying process which promotes wrinkles.

Essential oils will readily diffuse into the epidermis because of their lipophilic nature (long carbon chains). However, the stratum corneum with its layers of dead cells limits the rate of absorption to the rate of diffusion, as there is no circulation of fluids which could carry the oils deeper into the skin. Thus, where the stratum corneum is thinnest, the essential oils will get in fastest. For aromatherapists who are using the oils on a regular basis, it is somewhat of a comfort to know that the hands have relatively thick epidermis, so that the oils will preferentially go into the thinner skin of your client's body.

The main hazard to aromatherapists is the possibility of developing skin-sensitisation reactions, (or worse), simply by the repeated use of the oils over long periods of time, particularly if treatments are given several times a day, every day. I know one aromatherapist who now can't touch anything which has Lavender oil in it, even soap or handcream, and who also has developed sensitivity to some of the other oils such as Jasmine. If she does, she has an out-break of a weepy dermatitis which takes a while to heal.

In the living epidermal cells, there are enzymes of the same family as the liver cytochrome P450 enzymes[77] that begin the process of detoxification of lipophilic substances that manage to penetrate the stratum corneum. This includes a detoxification of the essential oils, which appear to be

---

[77] Rawlins MD (1989) "Clinical Pharmacology of the Skin" in *Recent Advances in Clinical and Pharmacology and Toxicology* Ed. Paul Turner, Churchill Livingstone, Edinburgh pp. 121-135.

potential toxins to the epidermal cells, merely because they are foreign. The chemical changes include the formation of C=C double bonds, the addition of -OH groups, forming alcohols, and the addition of water-soluble molecules to the essential oil molecules in a reaction called conjugation.

Dermis

This layer is 2 - 4 mm thick, and contains blood and lymph vessels, and nerve endings which are responsible for sensation of temperature, pain and fine touch control. It also has glands and hair follicles embedded in it which derive from the reproductive layer of the epidermis. These glands are responsible for the production of sebum and sweat, two substances which are important for the proper functioning of the skin as a protective barrier organ.

Sebum contains about 60% glyceride and free fatty acids, 25% long carbon chain esters (waxes), 15% precursor molecules, from which vitamin D_3 is made when ultra-violet light strikes the skin. The main function of sebum is to reinforce the imperviousness of the keratinised cells to water, and to prevent them from becoming too brittle.

The free fatty acids are formed by bacterial action on lipids, which are present in healthy skin in large numbers. Sebaceous glands which make the sebum are usually associated with hair follicles, and thus do not occur on the palms and soles. The greatest concentration occurs on the forehead and on the scalp, and may be one reason why massaging the essential oils into the scalp is such an effective form of treatment (try an Indian scalp massage!). The lipophilic nature of the sebum allows the essential oils to penetrate the epidermis via the sebum-lined hair follicles.

Sweat is another substance produced and excreted by the skin, and contains mainly water, and other water soluble substances such as sodium, potassium, urea and glucose. Its main function is to cool the body down by evaporating, the heat required for evaporation causing the body to lose heat. However, there are other instances of sweat production, which are caused by mental and emotional stress, or by eating spicy foods. Certain illnesses such as nausea, fever, and alcohol consumption also induce sweating, as do drugs like aspirin.

Some essential oil constituents and other volatiles found in food will be excreted through the sweat. It can be noticed in the body odour of people who eat different cuisines. For example, people who eat an Indian diet tend to smell of tumeric, ginger, chilli and cardamom, whereas people who eat an English diet tend to smell of rancid butter, milk or cheese ("cheesy" feet!). People who eat a Japanese diet tend to smell of seaweed and fish, with overtones of fermented soy products, and those who eat the maize-based diet of most African countries smell earthy and sweet (according to me).

The main function of the dermis is to provide the elastic and cushioning effect of the skin. It contains mainly molecules which are secreted by cells, and hardly any cells at all, compared to the epidermis. These molecules are the proteins collagen (30%) and elastin (1%), which are held in place by inter-molecular reactions with polysaccharide molecules, forming a bed for the collagen and elastin to attach to.

This "bed" is very hydrophilic, and attracts many water molecules, so it forms a kind of gel, which is easily deformed when pressed, but reforms when released. This layer lets the body's "clean-up" white blood cells move freely through it, which is useful, particularly when the skin is wounded, as it allows the "ambulance service" white blood cells to get through quickly to the injured area.

Penetration of essential oils
The hydrophilic nature of the dermis means that the lipophilic essential oil constituents would pass preferentially into cell membranes or attach to the lipoprotein elements of the blood.

According to Dr Pénoël, essential oils penetrate the blood system within a few minutes of application to the skin's surface, but this is only the first few molecules. For all the oil to penetrate through the skin, it takes a long time, as the process of diffusion is one which gets slower as it reaches the end point. It also depends on the size of the molecules, as to whether they can easily pass along the matrix of cell membranes.

Most essential oil constituents take from 20 - 60 minutes to completely penetrate the skin, and as a rule of thumb, monoterpenoids penetrate more quickly than sesquiterpenoids. 1,8-cineole and alpha-pinene penetrate within the first 20 minutes, whereas linalool, eugenol, and anethole take 20-40 minutes, methyl salicylate 40-60 minutes, cinnamaldehyde 60-80 minutes, and geraniol and citral take 100-120 minutes (Pénoël & Franchomme, 1990, p.297). Dr Pénoël collected these times from a variety of published articles, most of which were published in the 1940s and 1950s.

However, there is less interest these days in the actual penetration of the oils themselves, and more in their use as enhancers of penetration for other drugs. It is taken for granted that terpenoid molecules will to a greater or lesser extent, penetrate the skin.

Percutaneous absorption enhancement

A number of different terpenoids are being examined as carriers for trans-dermal drugs. For example, Zhao and Singh (1998) found that limonene, eugenol and menthone all aided the penetration of a larger molecule, tamoxifen through the stratum corneum. It seems they do so by partially dissolving the lipids in the stratum corneum, and making the stratum corneum lipids the preferable solvent for the tamoxifen[78].

Williams and Barry (1991) examined different types of terpenes as skin penetration enhancers for a polar molecule 5-fluorouracil. Their initial hypothesis was that the terpenes would "chase" the drug into the membranes of the stratum corneum (i.e. increase the partitioning of the drug into the membranes), but they decided that the terpenes were affecting the lipids between the cells of the stratum corneum, and thus allowing the drug to penetrate between the cells[79].

To my mind, this has implications for essential oil molecules being their own penetration enhancers. Interestingly, Williams and Barry found that 1,8-cineole was the most effective terpenoid at increasing penetration of the polar (i.e. non-lipophilic) drug (a 95-fold increase). This is not to say that we should use 1,8-cineole (or Eucalyptus oil) as an ingredient in all our blends, but it may be worthy of further exploration.

In a further study on sesquiterpenes, Cornwell and Barry (1994) found that sesquiterpenols or multi-functional sesquiterpenoids with at least one polar hydroxyl group were more effective as enhancers for the penetration of the same drug, 5 fluorouracil. Nerolidol was the most effective, causing a 20-fold increase in penentration (not as effective seemingly as the 1,8-cineole from the previous study). Another interesting discovery was that the

---

[78] Zhao K & Singh J (1998) "Mechanisms of percutaneous absorption of tamoxifen by terpenes: eugenol, d-limonene and menthone" *Journal of Controlled Release* 55(2-3) pp. 253-260.

[79] Williams AC, Barry BW (1991) "Terpenes and the lipid-protein-partitioning theory of skin penetration enhancement."*Pharmcological Research*.8(1) pp. 17-24.

sesquiterpene effects lasted for about 4 days, and that the sesquiterpenes would not wash out of the skin, even with a 50% alcohol rinse[80].

Godwin and Michniak (1999)[81] found that terpinen-4-ol and alpha-terpineol enhanced the penetration of hydrocortisone 3.9-fold and 5-fold respectively, which may have implication for use of essential oils such as *Melaleuca alternifolia* with topical cortisone creams. However, a caution is necessary here, as the experiments were done on hairless mice, and may not have the same effects in humans.

Menthol is another monoterpenol studied for its effects on increasing permeability of the stratum corneum to drugs. Kaplun-Frischoff and Touitou (1997) found that when menthol is mixed with the lipophilic testosterone, it significantly lowers the melting point of testosterone (normally a solid at room temperature). It also increases its solubility in ethanol, and combined with the effects menthol has on the stratum corneum lipids, it caused an 8-fold increase in permeability to testosterone[82].

Topical anti-inflammatory effects
Essential oils have some of their most poignant effects at the level of the dermis and epidermis, namely their anti-inflammatory effects. Dermal inflammation has three main stages[83]:

- vaso-dilation, increased permeability of blood vessels and release of pro-inflammatory agents
- migration of white blood cells to the damaged area
- repair of any damage

---

[80] Cornwell PA, Barry BW (1994) "Sesquiterpene components of volatile oils as skin penetration enhancers for the hydrophilic permeant 5-fluorouracil." *Journal of Pharmacy & Pharmacology.*46(4) pp. 261-269.

[81] Godwin DA, Michniak BB. (1999) "Influence of drug lipophilicity on terpenes as transdermal penetration enhancers." *Drug Development and Industrial Pharmcy.* 25(8) pp. 905-915.

[82] Kaplun-Frischoff Y, Touitou E. (1997) "Testosterone skin permeation enhancement by menthol through formation of eutectic with drug and interaction with skin lipids."Journal of Pharmaceutical Science.86(12) pp. 1394-1399.

[83] This section on inflammation is largely derived from Tortora GJ & Grabowski SR (1993) "*Principles of Anatomy and Physiology*" HarperCollins College Publishers pp. 695-697.

As mentioned previously in Chapters Three and Four, sesquiterpenoid molecules in particular seem to be able to inhibit inflammation. Their intervention seems to be at the level of preventing the release and production of the pro-inflammatory agents, and possibly the counteraction of the vaso-dilation by local effects on the smooth muscle of capillary walls.

Below are some of the pro-inflammatory molecules produced by the body:
- *Histamine*. This is a molecule containe in cells in the dermal layer, in particular the white cells known as mast cells. If any of the cells containing histamine get damaged, they release the histamine, which acts as the signal for initiating the rest of the inflammatory response. If you can prevent histamine release, you can prevent inflammation by preventing vasodilation, as all sufferers of hay-fever are only too well aware.
- *Kinins*. These molecules are normally found in the blood in an inactive form, and summon the white blood cells known as phagocytes to come and fight infection. Phagocytes engulf bacteria, disable them and die, to become pus if enough of them accumulate. Kinins also work with histamine to cause vasodilation and permeability of capillaries, which brings more blood to the area and increases heat, redness and swelling. Kinins cause much of the sensation of pain by irritating the pain receptors in the dermis.
- *Prostaglandins*. There are 3 main types of prostaglandin, E1, E2 and E3 (often written PGE1, PGE2 etc). PGE2 is the pro-inflammatory prostaglandin, whereas PGE1 and PGE3 actually work in an anti-inflammatory manner. Prostaglandin E2 intensifies the effects of histamine and the kinins, and may summon the phagocytes to the area also. It is released by damaged cells in the dermal layer, and has the effect of prolonging and intensifying pain caused by stimulation of the pain receptors by the kinin molecules.
- *Leukotrienes*. These small molecules signal the site of infection and damage for the phagocytes, and are produced by basophils and mast cells.
- *Complement proteins*. These proteins are in the blood, and are necessary for the destruction of bacteria at the site of infection.

Both prostaglandins and leukotrienes are eicosanoid molecules, which are formed from gamma-linolenic acid, the beneficial fatty acid in Evening Primrose oil. Formation of the pro-inflammatory PGE2 is controlled by the presence of insulin in the blood, which activates the formation of arachidonic acid by the enzyme delta-5-desaturase. Arachidonic acid is the precursor of PGE2 and the leukotrienes, and is itself toxic enough to kill animals when injected into them (Sears and Lawren, 1995)[84].

In order to prevent formation of PGE2, Sears and Lawren recommend that you keep insulin levels low, eat foods containing eicosapentanoic acid such as salmon and some other fish (as this molecule inhibits the activity of delta-5-desaturase), and ensure that your body has adequate levels of gamma-linolenic acid in the first place, by eating enough, reducing your stress and avoiding viral infections. The whole book, "The Zone: A dietary road map" makes fascinating reading, and presents the story of how eicosanoids are integral to health and well-being in easy enough terms to be understood without a medical background.

Below are the diagrams of gamma-linolenic acid, arachidonic acid and prostaglandin E2[85].

gamma-linolenic acid

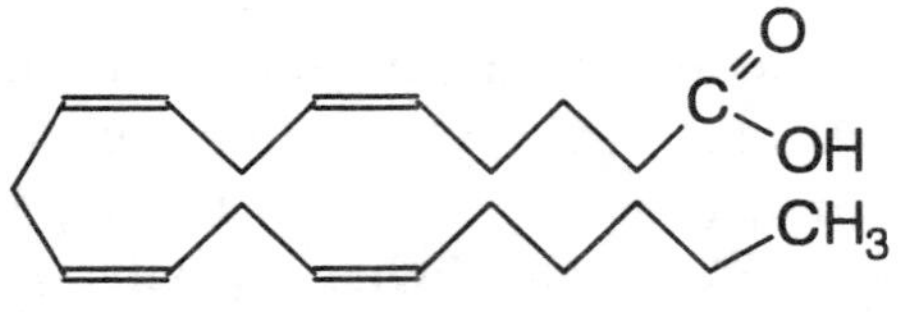

arachidonic acid

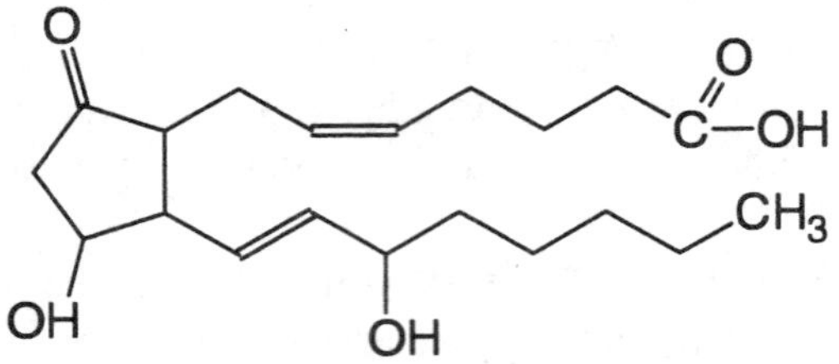

prostaglandin E2

[84] Sears, B and Lawren W (1995) "*The Zone: A dietary road map*" Harper Collins, NY, p126.
[85] Lehninger A (1982) "Principles of Biochemistry" Worth Publishers, NY, p. 747.

It is worth considering the structures of the molecules, as there are certain similarities between the sesquiterpenoid and coumarin structures and these molecules, which may indicate some possible interaction. Aspirin (acetylsalicylic acid) works as an anti-inflammatory agent by preventing the formation of prostaglandin E2 from arachidonic acid. Yin et al. (1998) have determined that aspirin manages this by inhibiting activation of NF-kappaB transcription factors which code for the production of the prostaglandin synthase enzymes and other enzymes needed to create the inflammatory response[86]. Lyss et al, (1997) as mentioned in Chapter Four, found that helenalin, a sesquiterpenoid lactone found in Arnica extracts also inhibits the NF-kappaB transcription factors, but by a different mechanism than aspirin does[87]. The structure of helenalin, a sesquiterpenoid lactone (refer to Chapter Four, section on lactones for details), and that of aspirin and methyl salicylate (found in Wintergreen and Birch essential oils) are shown below:

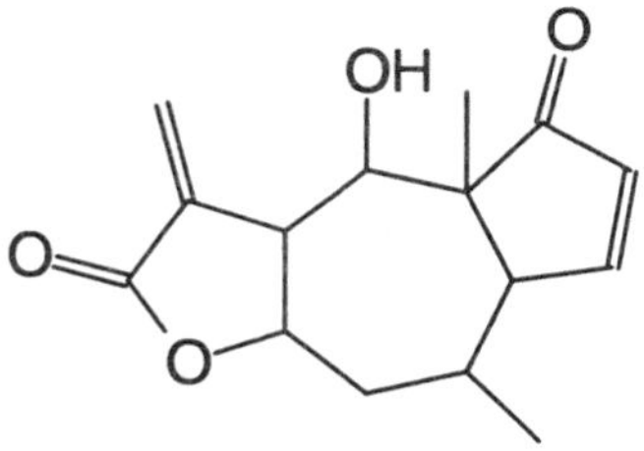

helenalin

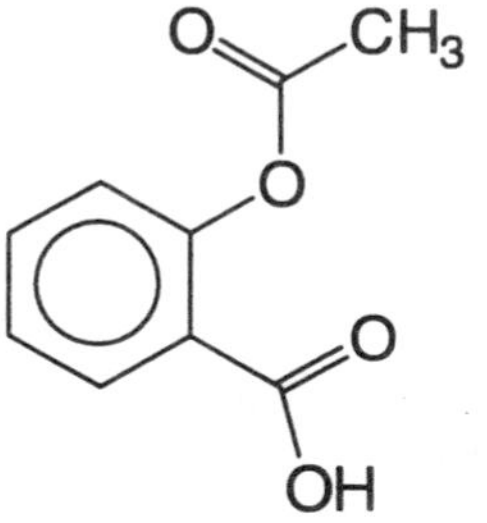

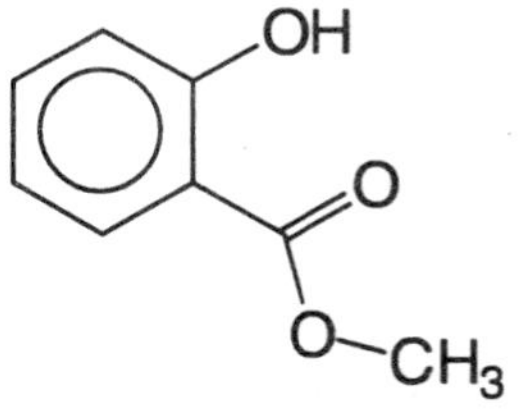

acetylsalicylic acid
(aspirin)

methyl salicylate
(wintergreen and birch oils)

---

[86] Yin, MJ, Yamamoto Y, Gaynor RB (1998) "The anti-inflammatory agents aspirin and salicylate inhibit the activity of I(kappa)Bkinase-beta." *Nature* November 5 396(6706) pp. 77-80.

[87] Lyss G, Schmidt TJ, Merfort I, Pahl HL (1997) "Helenalin, an anti-inflammatory sesquiterpene lactone from Arnica, selectively inhibits transcription factor NF-kappaB." *Biological Chemistry* 378(9) pp. 951-961.

Subcutaneous layer (adipose layer)

This layer is made up of fat cells and provides the insulation required for maintaining body temperature at a constant level. It is also where the body stores most of its excess energy acquired from digestion of food. The essential oils do not penetrate through the skin to this level, but parts of them may end up stored in this layer after the body has metabolised them, due to their lipophilic nature.

RESPIRATORY SYSTEM

The use of the respiratory system as a method of getting the oils into the body depends on an equilibrium being achieved between the amount of essential oil vapour in the air which the lungs breathe in, and the amount of essential oil actually absorbed by the alveoli of the lungs. It would seem that inhalation of essential oils is good way to get them into the system, as there is no stratum corneum preventing their entry.

Falk et al. (1990) noted that uptake of alpha-pinene in human volunteers exposed to 450mg/m^3 of alpha-pinene in the air was about 60%. About 8% of the alpha-pinene was exhaled after the exposure was stopped, and the rest of it was removed from the body in the urine. The rate of clearance from the blood was high, but when the poorly perfused tissues such as adipose tissue were examined, there was a long half-life of the alpha-pinene (about 30 hours) in these tissues. This implies that the alpha-pinene is in fact stored in the adipose tissues for a few days[88].

Inhalation is obviously the most effective application method for topical treatments of respiratory problems such as bronchitis and other infections, although there is some equivocal information about use of essential oils for asthma as noted in Chapters Three and Four (Monoterpenes, Ketones and Lactones).

METABOLISM OF OILS IN THE BODY

When the essential oils have got into the blood stream from the skin, the circulation directs them to the heart, then the lungs, then the heart again,

---

[88] Falk AA, Hagberg MT, Lof AE, Wigaeus-Hjelm EM, Wang ZP (1990) "Uptake, distribution and elimination of alpha-pinene in man after exposure by inhalation" *Scandinavian Journal of Work Environment and Health* 16(5) pp. 372-378.

and then over the rest of the body. This is an advantage over taking the oils internally, as by that method, the oils are metabolised by the liver before being circulated throughout the body. What follows is a brief overview of the types of reactions that happen to essential oil constituents when they pass through the liver and other organs. There is enough additional information available on the pharmacokinetics and metabolism of individual constituents to fill another book, but I am aiming here for a foundational understanding rather than detail.

Liver

The liver is the main centre for the breakdown of essential oils which in terms of the body's internal environment, are foreign materials (xenobiotics). The liver responds by transforming them into something more useful, or into something that can be excreted. The cytochrome P450 enzymes work on the essential oil constituents to make them either:

- more water-soluble, by the addition of a sulfate or glucuronate group, or the formation of an alcohol or acid, in readiness for excretion by the kidneys
- more simple, so that the body can use the smaller component parts (e.g. acetyl groups) in its ordinary reactions. This releases the energy stored in the bonds of the essential oil constituents, and is part of the way in which the liver normally deals with lipids.

For more chemical details of metabolic reactions, please see my companion book "The Essential Oil Chemistry Workbook" (1995, self-published).

Liver toxicity

Sometimes the metabolic products can be more harmful than their parent constituents. An example of this is the ether safrole, found in Sassafras oil (90%), and less than 1% in oils like Cinnamon, Nutmeg, Star Anise and Ylang ylang. In rats, the metabolism of safrole gives an epoxy compound which is causes liver cancer.

Fortunately, humans have a different liver enzyme, which yields an acidic form rather than the epoxide form, although it is not known whether the acid may or may not be harmful (Pénoël & Franchomme, 1990, p. 178; Tisserand & Balacs, 1995, p. 198). Another example, also mentioned in Chapter Four with the ketones is the formation of menthofuran from the ketone pulegone, which destroys cytochrome P450 enzymes in the liver cells, thus preventing them from being able to detoxify other substances.

As Tisserand & Balacs (1995, p. 61) so clearly discuss, another molecule in the liver which is essential for the removal of free radicals and other toxic substances, is glutathione. Quite a bit of work has been done on the effect of essential oil constituents on glutathione and glutathione-S-transferase, and the following molecules taken internally lower the liver's levels of glutathione: trans-anethole, cinnamaldehyde, methyl chavicol (estragole), eugenol, methyl eugenol, d-pulegone, safrole, and possibly apiole (Parsley oil). With the exception of d-pulegone and cinnamaldehyde, they are all phenyl methyl ethers. I would hazard a guess at saying that all phenyl methyl ethers, and probably most molecules with the aromatic benzene ring in their structure would be likely to deplete glutathione levels in the liver.

In the event of poisoning with liver toxic essential oils, it would be vital to call a Poisons Centre as well as the ambulance. One piece of information that has come my way is the treatment of paracetamol or acetaminophen (see diagram below) overdose with a substance called N-acetyl-L-cysteine, which is a modified amino acid. It acts as a glutathione substitute, and also has been shown to chelate copper and zinc in animal experiments. If no other treatment is suggested, it may be a possible option for treatment of poisoning with oils that deplete glutathione levels in the same way as paracetamol.

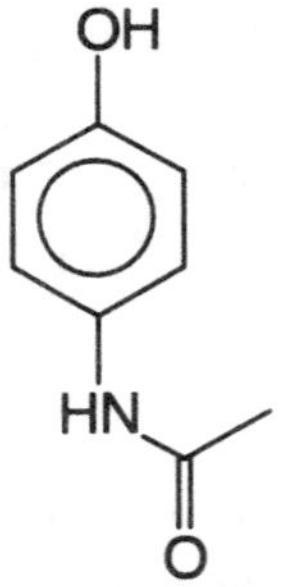

acetaminophen (paracetamol)

Interactions of essential oils with pharmaceutical drugs
The main caution according to Tisserand & Balacs (1995, p. 61) is not to use essential oils with glutathione-depleting molecules in them whilst taking paracetamol or other drugs which may contain acetaminophen. This is because paracetamol also needs glutathione to be metabolised, and if there is not enough to go round, the paracetamol may cause liver damage. The same caution goes for people who have liver diseases such as cirrhosis or hepatitis. The reader is referred to Tisserand & Balacs

"Essential Oil Safety" (1995, Churchill Livingstone, Edinburgh), Chapter 4 on Metabolism for further reading on this topic.

Heart

Essential oils may affect the heart by their following actions:

- causing vaso-dilation or vaso-constriction which alters blood pressure, and therefore decreases or increases the load on the heart muscle. Tisserand & Balacs (1995, pp. 65,66) suggest that ingestion of large amounts of l-menthol (e.g. eating a whole packet of peppermint sweets, or smoking mentholated cigarettes) can cause cardiac fibrillation (the heart losing its normal rhythm).

- interacting with adrenaline receptors producing either tachycardia (speeding up the heart-beat) or bradycardia (slowing it down). During toxicity reactions, tachycardia is a symptom often noted, particularly with camphor, 1,8-cineole and methyl salicylate (Tisserand & Balacs, 1995, pp.50-57).

Brain

Due to the small size and lipophilic nature of the essential oil constituents, it is most likely that they cross the blood-brain barrier (monoterpenes are smaller than caffeine molecules). Whether they are actually taken up by the olfactory neurons from inhaled air and transported to the brain is a moot point.

The ketones and ethers are most likely to have an deleterious effect on the brain, as shown in Chapter Four. Large oral doses of ketones may have implications for the demyelination of neurons, which would be harmful to say the least. The so-called "narcotic" effects of some essential oils may in fact be the symptoms of mild poisoning. A comparison with the effects of drinking alcohol is relevant here - people like the numbing, relaxing feeling, but it is actually a sign of mild poisoning.

On the other hand, sesquiterpenols like eudesmol have been shown to prevent artifically induced seizures (by drugs and electroshocks) (see Chapter Four).

Kidneys

The kidneys are the major route for excretion of the essential oil metabolites, but they can be excreted via the skin, lungs and digestive tract too. Non-polar substances will not be excreted by the kidneys, as they are too readily reabsorbed across the filter membranes. However, polar substances will not easily pass back across the filter membranes, due to extra energy requirements for them to do so, so they will be excreted as urine. Tisserand & Balacs do not mention any particular essential oil constituents which are toxic for the kidneys, with exception of apiole found in Parsley oils (leaf and seed). However, due to the amounts safely consumed in a dish of tabouleh, or even used in a 2.5% blend aromatherapy treatment, it is unlikely that Parsley oil would present any problems.

NEUROENDOCRINE SYSTEM

First of all, we must look at the nervous system and the integration of the endocrine system, before looking at the effects which the essential oils have on them both. The nervous system is made up of two main parts, the somatic and the autonomic pathways. The somatic part is under voluntary control, whereas the autonomic is for the most part involuntary. Both somatic and autonomic pathways have neurons in the central nervous system (CNS), and in the peripheral nervous system (PNS). The brain and the spinal cord compose the CNS, and all other nerves are part of the peripheral system.

The autonomic pathway can be subdivided into two parts, which have generally opposite effects on the body. The parasympathetic pathway is concerned primarily with getting the body to "rest and ruminate". The other is the sympathetic pathway, and is concerned with getting the body ready for "fight or flight". The neurons of each pathway are positioned differently, and also release different substances as their transmitters of information. This difference lets the target organs know which response to make. In general, adrenaline is the stimulant for the sympathetic receptors on neurons, whereas acetylcholine stimulates the parasympathetic ones.

The nervous system monitors both the outside and internal environment of the body, and as such, is constantly receiving and transmitting information via nerves to and from the CNS, and the PNS. Most of the information which is received will cause the endocrine system to be activated, and hormones will be produced which carry specific information to target sites throughout the body. The main centre for hormonal control is the hypothalamus, in the brain. The pituitary gland and the hypophysis are also concerned with hormone regulation. Hormones released from these control glands travel throughout the body until they reach their target, usually another endocrine gland, and further hormones are released.

A significant example of how nerves and hormones work together is the response of the body to a perceived stress (like a car pulling out in front of you suddenly). The hypothalamus receives messages from the peripheral system (receptors in the eyes and ears), and information from the cortex (thinking part) of the brain, which indicate stress. One of the hormones which is released stimulates the pituitary gland to release another hormone. This one goes straight for the adrenal glands on top of the kidneys.

When it binds to the appropriate receptors in the adrenal gland, it stimulates the production of adrenaline (epinephrine) which is absorbed into the blood. Adrenaline stimulates the sympathetic neurons of the autonomic nervous system, which then cause the following effects:
- dilating the pupils,
- raising blood pressure by vaso-constriction,
- quickening the heart beat,
- causing the breathing to bcome deeper and more rapid,
- increasing the production of glucose by the liver and muscles to prepare for rapid muscular response (e.g. jump out of the way, slam on the brakes or swerve),
- slowing down digestive processes[89].

How essential oils affect the neuroendocrine system

There are two main ways in which essential oils affect the neuroendocrine system:
- via the olfactory route
- via stimulation of nerves and glands as a result of being in the blood.

The olfactory route is interesting to me, particularly as it relates to modification of mood and behaviour, and may also have some indirect effect on structures like the hypothalamus. Odourous substances are breathed in through the nose, dissolve in the olfactory mucosa and may stimulate the olfactory nerves which have fine hair-like endings called cilia embedded in the mucus. They may also stimulate the trigeminal

---

[89] Further information on hormones and nerves can be found in Tortora GJ & Grabowski SR (1993) *"Principles of Anatomy and Physiology"* HarperCollins College Publishers, Chapter 18, pp. 556-560.

nerve, which is a branched nerve responsible for detecting pain, touch and temperature in many parts of the face and head including the nasal epithelium. Constituents such as menthol and 1,8-cineole cause their sense of cooling and clearing via their interaction with the trigeminal nerve endings in the nose.

Very small molecules (less than four or five atoms) and very large molecules (over 100 atoms) generally have very little odour, but they do sometimes have an effect on odour perception, either by filling receptor sites, or by interacting with other odorant molecules. Amoore has long been a proponent of the stereochemical theory of olfaction, and his theory still stands today, although there are modifications to it. He suggests that there are olfactory receptor sites on the cilia of the neurons which respond to particular shapes of molecules, and that there are only a few major types of receptors (Amoore, 1982)[90].

It has also been found that a particular molecule is identifiable not only by the way it reacts with olfactory nerve cell receptors, but also with the pattern with which several molecules of the substance affect the whole olfactory epithelium (Shepherd, 1991)[91]. Information patterns are collected in the olfactory bulb, and from there are relayed to various areas of the brain, including the amygdala, hippocampus, hypothalamus and the anterior part of the pituitary gland. The amygdala is responsible for fundamental mood states such as rage and fear, whereas the hippocampus is associated with laying down long-term memory patterns. It is thought that the reason why smells can so powerfully trigger memories of childhood is due to the close association of the olfactory tracts with the hippocampus and memory-making mechanisms.

Although it has not been established what exact effect essential oil molecules have on the olfactory system, there have been studies done which show that substances taken in via the nose, do have effects on both brain wave patterns and on other monitors of nerve stimulation, such as heart rate and blood pressure [92].

---

[90] Amoore JE (1982) "Odour Theory and Odour Classification", Chapter 2, *Fragrance Chemistry: The science of the sense of smell* Ed. ET Theimer
[91] Shepherd GM (1991) "Computational Structure of the Olfactory System", Chapter 1, *Olfaction: A model system for computational neuroscience* Eds. JL Davis & H Eichenbaum, MIT Press, USA
[92] van Toller S & Dodd GH (1991) "*Perfumery, the psychology and biology of fragrance.*" Chapman & Hill

Certainly, the sense of smell is one of the most primitive which we have, and it is definitely linked to that part of the brain which is purported to deal with our basal emotions (the limbic system). The sense of smell is also supposed to be linked to our choice of sexual partner - we will be attracted to someone who emits the right smell, though the "right" smell, whether pheremonal or artificial, is bound to be highly subjective (Lake, 1989)[93].

Obviously, a great deal of work still needs to be done in the area of how essential oils affect the neuroendocrine system, in particular via the olfactory pathway. However, we know that essential oils do work, and so as long as they are used wisely, it is a matter of waiting for science to catch up with history.

Systemic effects of essential oils on nerves and hormones

The majority of work which has been done on the effect of oils on the nervous system relates to either the toxic effects, or the topical effects related to their anti-inflammatory effects. As mentioned in Chapter Four, some ketones have a damaging effect on nerves, including causing CNS effects such as seizures, dizziness, and blurred vision. The oxide 1,8-cineole at toxic dosages also causes similar effects.

Menthol, geraniol, alpha-terpineol and linalool have mild anaesthetic effects according to Dr. Pénoël (Pénoël & Franchomme, 1990, p.158), and he suggests that the monoterpenols in general also have a vasoconstrictive effect, presumably on the dermal capillaries during a topical application. Of the mechanisms which have been researched, there is evidence that some essential oils (no individual work done on the constituents) prevent the uptake of calcium Ca^{2+} ions at the neuromuscular junction in smooth muscle, or block neuromuscular transmission, thus preventing muscular contraction (Hills & Aaronson, 1991[94]; Zygmunt et al., 1993[95]; Albuquerque et al., 1995[96]).

---

[93] Lake M (1989) *"Scents and sensuality: The essence of excitement"* John Murray (Publishers), London, pp.3-19.

[94] Hills JM, Aaronson PI (1991) "The mechanism of action of peppermint oil on gastrointestinal smooth muscle. An analysis using patch clamp electrophysiology and isolated tissue pharmacology in rabbit and guinea pig" *Gastroenterology* 10(1) pp.55-65

[95] Zygmunt PM, Larsson B, Sterner O, Vinge E, Hogestatt ED (1993) "Calcium antagonistic properties of the sesquiterpene T-cadinol and related substances: structure-activity studies" *Pharmacology & Toxicology* 73(1) pp.3-9

[96] Albuquerque AA, Sorenson AL, Leal-Cardoso JH (1995) "Effects of essential oil

As for hormonal effects, there has been no molecular biochemistry done on the exact mechanisms of action, but there are suggestions that essential oils may have some effect on the steroidal hormones, either competing with them for binding sites, or inhibiting their production by affecting the responsible gland. These suggestions are mainly from Dr. Pénoël, and seem to be based on the structure-activity hypotheses, which are yet to be verified *in vivo* or *in vitro*.

IMMUNE SYSTEM

The immune response is another complicated system, which though not directly under the control of the neuroendocrine system, nevertheless utilises chemical messengers, and can be augmented by the neuroendocrine system.

The immune response is initiated by the invasion of the body by foreign matter, or by a breaching of the skin, for example, when it is wounded. Injured cells send out the first messages, histamines, which have the effect of dilating the blood vessels and allowing more blood to come to the area (redness and swelling). The clean-up crew, in the form of white blood cells, are called to the scene, to clear up any invaders or wreckage of self.

If it is an invasion of other life-forms, such as bacteria, or viruses, then a second line of defence is called in. This is the anti-body system, which is ordered by other white blood cells which are mainly found in the lymph nodes. When they receive the messages from the wounded cells, some of the white blood cells (lymphocytes) come on the scene. Instead of cleaning up the debris, they latch on to any foreign thing, and take an imprint of it, or the actual thing, and take it back to the lymph nodes. Once there, the lymphocyte starts to divide, and its daughter cells are the next step in the defence. (This accounts for the swelling of the lymph glands during infection).

Most of the daughter cells become factories for antibodies (immunoglobulins), which are molecules designed to exactly fit the bit of the intruder which their parent cell captured. These antibodies are released into the blood stream, and attach themselves to whatever they find which they match. An intruder, thus attached with antibodies is

of Croton zehntneri, and of anethole and estragole on skeletal muscles" *Journal of Ethnopharmacology* 49(1) pp.41-49 marked, and other clean-up cells can easily recognise it and come and clean it up.

The system is much more complicated than this, with some lymphocytes being required for stimulation of the clean-up cells, and it is when these lymphocytes, the helper cells, are infected, that auto-immune deficiency syndromes can be initiated. It is thought that the Human Immune system Virus (HIV) is one that attacks the helper cells, thus preventing them from helping the clean-up cells do their job. As a result, a victim of this virus will become susceptible to infection from virtually every type of disease, which their normally healthy body and immune system would have destroyed. This is where the anti-bacterial and anti-fungal properties of the essential oils can come in handy, as they work powerfully with few side effects if used in aromatherapy doses. Several palliative care centers and nursing homes vaporise essential oils not only for mood modification, but for disinfectant purposes as well.

How essential oils affect the immune system

Very little has been actually proved as to whether essential oils can stimulate the immune system, in cases such as AIDS, or cancer. However, if the body is in good working order, and if all the other systems are in harmony, then there is the probability that the body will be able to hold out longer against disease. And it has been shown that the essential oils for the most part can be used to get the body back into equilibrium, so I would say that this is the major effect of the essential oils on the immune system - an indirect one.

However, it may be found that certain of the constituents imitate chemical messengers which the body puts out during invasion, and that they may be instrumental in promoting the antibody response. According to Dr. Pénoël, experiments have been done which suggest the amount of immunoglobulin in the blood changes after treatment with essential oils (Pénoël & Franchomme, 1990, p.187-188), but the results are unclear and await further research.

SUMMARY OF THERAPEUTIC PROPERTIES OF CONSTITUENTS

The following list is a set of generalisations, regarding whether a constituent is "calming" or "stimulating". According to Dr Pénoël, a constituent will fit in a particular category according to the reaction of its ionic form in an electric field (the quadrant I mentioned earlier). Calming constituents form negative ions, tonifying stimulants form positive ions in an electric field. The therapeutic properties have been hypothesised and tested by Dr Pénoël in France (Pénoël & Franchomme, 1990, pp. 99-100).

Calming and Relaxing Constituents

Coumarins	Sedative; anti-lymphoedemic
Esters	Anti-spasmodic; sedative; immuno-modulant (?)
Ketones	Mucolytic; wound healing; neurotoxic (in larger doses)
Lactones	Mucolytic and expectorant
Sesquiterpenes	Anti-inflammatory; anti-histaminic; anti-allergic
Aldehydes	Anti-infectious (medium strength); anti-inflammatory

Stimulating and Toning Constituents

Aromatic aldehydes	(eg. cinnamaldehyde) Anti-infectious (strong).
Phenols	Anti-infectious (strong); skin and mucous membrane irritants
Ethers	Anti-spasmodic (for gut mainly); analgesic (some); anti-infectious ("all or nothing" effect)
Oxides	Expectorant (strong); anti-parasite
Monoterpenes	General tonic (decongestant); adrenalin stimulant (some); muscular aches (some); antiseptic
Monoterpenols	Anti-infectious (strong); vaso-constrictors/local anaesthetics (some); immuno-stimulants (?)
Sesquiterpenes	Can also be stimulating, with similar properties
Sesquiterpenols	Tonic; estrogen-like (some ?)

The main purpose of trying to assign therapeutic properties to functional groups is to make it easier to grasp the chemistry. However, I think that in the end it is more useful to examine each molecule in its own right, as sometimes you will come across a molecule which has contradictory properties for its functional group. An example is the monoterpenol linalool, found in many essential oils. According to the above list, you would expect it to be in the stimulant category, but it has been discovered to be a sedative to the central nervous system.

Another hazard is that the majority of research is performed on animals, and at doses much higher than we would ever use in aromatherapy, even the internal dosages used by European practitioners. Until there is some more research done on human volunteers, it is a game of guess-work and intuitive matching of practical experience with scattered pieces of research information as I have done in this book.

CHAPTER SIX

QUALITY CONTROL OF ESSENTIAL OILS

Having seen that the essential oils do affect the body, and in particular that it depends on their chemical constituents as to what effects they have, it is obviously important to think about quality control. In 1993 I had the opportunity to work as quality control manager in a large essential oil importing company in Australia. I learned a great deal about the essential oil industry, and noted that most essential oils are not produced for the delight of aromatherapists, but for the flavour and fragrance industry. As such, this is a personal reflection of the things I learned about quality control during my time with that company.

In the search for "pure and natural" essential oils, the following three issues are the most likely to cause aromatherapists concern, particularly when considering such desirable properties as life force and synergy.

Degradation of essential oils

So far we have investigated the types of essential oil constituents, and with most of them, we have seen that they are reactive compounds, which will readily combine with oxygen from the air, especially if there is any free energy around in the form of heat or light. Some will form resins (polyterpenes), and others will be oxidised, which is the other main reaction which takes place in the essential oils. There will be some rearrangement of the positioning of double bonds, and even sometimes of the positioning of the functional group. These processes may alter the medicinal properties of the essential oil constituents, so it is important to store the essential oils in conditions which:

- Minimise contact with the air. Small bottles with narrow necks and valve-like stoppers, or dropper bottles are best.
- Minimise contact with free energy. Dark-coloured glass bottles and refrigeration of the citrus and other mainly monoterpene oils prolong the shelf-life of the oils.

Another precaution is to not keep essential oils for more than about three years, as they do start to rearrange and degrade. However, no definitive studies have been done on the length of time for significant degradation of constituents, or on the effects of such degradation on the therapeutic qualities of the oils. In the case of patchouli oil, it is considered that the older the oil is, the more efficacious it is, possibly because the aroma is richer and more rounded, but also probably because the oxidation of the molecules has in this case made them more potent therapeutically.

Extraction of oils from plants

There are five main methods of extraction:

- steam distillation
- cold-pressing (for citrus oils only)
- extraction with chemical solvents
- enfleurage, whereby the plant material is pressed between two layers of wax until all lipophilic materials have been absorbed
- super-critically heated and pressurised carbon dioxide (often known as CO_2 extraction).

The super-critical CO_2 extraction technique extracts most of the constituents from the plant material under pressure into the carbon dioxide, which then evaporates off, leaving you with a pure extract. The heat used is lower than that of steam distillation, which prevents the formation of certain "products of distillation" which are well known, and sometimes sought after in the fragrance and flavour industry (for example, the distinctive odour of lime oxide in steam distilled Lime is not present in cold-pressed lime oil).

A CO_2 extract contains all the molecules found in the essential oil, and often some molecules which would normally only be found in solvent extracts, such as flavanoids, chlorophylls/pigments, and a higher proportion of sesquiterpenoids and waxy di- and triterpenoids.

This would also be the case for extracts which are created by use of solvents which rely on sub-zero temperatures, like the phytols or florasols made by Wilde & Co., England.

As we have seen the sesquiterpenoid compounds have some potent properties. Increasing the sesquiterpenoid content could therefore alter the therapeutic properties possibly making CO_2 extracts more potent than essential oils, and maybe more hazardous too (think of the lactones). The only way to determine whether there is any difference between the CO_2 extracts and essential oils is to use them, and then compare their constituents with gas chromatographic analysis

Life force

In wondering what happens to the life force during these procedures, it is necessary to ask the question "Where does the life force reside in an essential oil?". If it is only associated with living matter, such as plant cells or animal cells, then four out of the five methods of extraction will definitely destroy the life force. Plant cells will be killed during steam distillation, as their cell walls are broken down, and as the temperature gets too high for the reactions to continue. Chemical solvents will prevent the cellular reactions from continuing in the cells, thereby causing death. The same applies to the CO_2 method. During cold pressing, the cells will be ruptured and therefore killed.

With enfleurage, it may be argued that the plant material is not actually killed immediately, as it is well-known that plant material can continue to live for a few days, even whilst separated from its parent plant. Thus it may be argued that the plant cells are still alive, even while their essential oils are being encouraged out of them. However, after a few days separated from the plant, the flowers will die, and presumably the life-force will then be extinguished.

Effects of extraction method on oil quality

Steam distillation

This method results in the retrieval of C_{10} and C_{15} terpenoid compounds, which are insoluble in water, but are volatile enough to be driven off at the temperature of steam, which depends on whether a pressurised system is being used. Usually temperatures do not exceed 100° C.

One question which is not often asked is whether any degradation of the constituents occurs under such temperatures. The answer is "Yes, rearrangement, hydrolysis and artefact formation do occur". Distillers therefore try to minimise the length of time which the constituents are actually held at those high temperatures. Time in terms of minutes would be the optimum for distillation, but obviously this depends on the quantity of plant material, and the industrial economics of such procedures. The smaller the still, the better the quality of the essential oil, is a general rule of thumb.

Cold pressing

In some ways, cold pressing obtains an oil which is closest in odour to the original fruit. Unfortunately, only the citrus peel oils yield enough by this method to make it worthwhile. It is possible to envisage cold-pressing seeds, but the cost would probably be prohibitive, and the yield minimal.

Chemical solvents

This method is more flexible than steam distillation, as you can tailor-make your solvent, to ensure extraction of all the components you require, in particular the constituents with longer carbon chains (C_{20} and C_{30}). However, as this is a relatively new method, the medicinal properties of essential oils containing these compounds are not known. The method involves extraction of the constituents from the plant material, and then removing the solvent by evaporation techniques. The questions that remain are:

- whether the solvent is ever completely removed from the resultant essential oil
- whether the solvent changes the properties of the constituents during extraction.

Enfleurage

This method is used for extracting constituents which are either known to degrade in the previous techniques, or which are present in such small quantities that every last little bit must be captured, for example, jasmine. The disadvantage of this technique is that it is so time-consuming, and that only small quantities can be properly processed at one time. It makes the oils thus derived very expensive.

As for synergistic effects, obviously these will exist between constituents of essential oils, and may be disrupted by the presence of even small amounts of solvent, water or wax, so the emphasis is on obtaining essential oils from a reputable supplier, who takes care with their extraction processes, and has a stringent approach to quality control.

Adulteration

Since certain essential oils are scarce or expensive, the temptation of producers is to add synthetic constituents, or constituents from other plant sources to "stretch" the essential oil, and also to lower the cost of production. As far as life force goes, I think we can say that essential oils cannot contain life force, so synthetic constituents will not affect the oils by diminishing the life force. Synergy is another matter. Synthetic constituents cannot be guaranteed to be 100% free of their starting materials - the best you can get, if you still want to keep the price low is about 95-99% purity. This means that there will probably be a disruption of the synergy between constituents, due to the presence of the starting materials.

As for the effect of adding constituents derived from other plant sources, you run into the problem of synergy again. How much must you add to keep everything in balance? As we have seen, the fragrance of the real essential oil is derived from all of its constituents, even the ones which are only present in trace amounts, so if you "over-do" one of the constituents, you will alter the fragrance, and also the medicinal properties.

This is the major problem of reconstituted essential oils, in that they often leave out the constituents which are only present in small amounts, as the focus is on lowering the price of production, not on producing a replica of the real essential oil. This would in fact end up being a far more expensive process than the mere extraction of the oil from the plant.

How can we be sure we are not buying adulterated oils?
The cheapest way to do this is to train your own nose, by finding a reputable supplier, and by smelling the oils, thereby storing up a memory bank of smells. This takes a long time, and is further complicated by the variation of constituents in different chemotypes of plants, and in the growth variables which the source plants were subjected to.

The next method is to be aware of the extraction process required for each essential oil, and the amounts of oil available from the plant. If, as with rose and jasmine oils, each ounce of essential oil is derived from 180 pounds of blossoms, you can be sure that these oils will be very expensive, unless they have been adulterated. This information is available from perfumery and fragrance books, for example, J W Poucher's three volume set " Perfumes, Cosmetics and Soaps"[97].

The more expensive (but more precise) way of finding out which constituents are in any particular essential oil is to have a gas chromatograph taken of them. Gas chromatography (GC) is a method which uses the different volatilities and solubilities of the essential oil constituents in a gaseous phase, to separate them out, and to record their concentration as a percentage of the whole.

The most useful type of gas chromatography is where the chirality of the constituents is taken into account. Natural oils generally contain constituents of homogenous chirality, whereas synthetic constituents will be a mixture of chiralities. Chirality is connected with the arrangement of atoms in different isomers of the same compound, and often with the

---

[97] Poucher WA (1959) "*Perfumes, Cosmetics and Soaps*" Chapman Hall, London

way in which the isomers refract light differently, i.e. their optical activity (see Chapter 7).

Gas chromatography can be combined with mass spectrometry (MS), which is a method which analyses each constituent according to the way it fragments when bombarded with ions. When coupled with gas chromatography, this method allows you to pin-point each constituent as it is separated out, and compare its fragmentation pattern with a library, thereby determining the identity of the compound. Such an analysis would be a GC-MS analysis.

Without a coupled mass spectrograph, gas chromatography is less effective, as you first have to discover the retention times of each constituent in the oil so as to be able to understand which of the peaks is which constituent. This means that all the constituents have to have been elucidated first of all, and this is where the problem arises, as it is often difficult to separate, let alone synthesise the constituents, or to obtain them in large enough quantities to test them.

What can be done, though, is for a standard gas chromatograph (GC) to be made, and then each subsequent essential oil, of that type, can be compared to it. Thus you do not need to know what all the peaks represent, but are interested in the pattern, and whether it matches or not.

Below is an example of a GC-MS analysis of a solvent-extracted Jasmine[98]. The constituents have been identified by comparison with a computer data-base of mass spectrograph traces. This allows the constituent present at each peak to be identified, although some of the smaller peaks elude identification, regardless of how important they may be. To find constituents which are key odorants in a fragrance, the GC-MS system can have a dual column which opens to the outside air at the same time as the peak is drawn on the chart. This allows a perfumer or analyst to note which of the peaks carry which odour, and can be an even more powerful tool than the MS, as the human nose can detect some compounds at parts per billion.

On the following GC-MS print out, only some of the peaks have been identified, and they appear in the chart below it with their percentages.

---

[98] The jasmine oil came from Wilde & Co., UK, and the article I wrote about it, "Vital Phytols - just in time for the new millenium" appears in *Aromatherapy Today* (1998) v8 pp. 38-41. Many thanks to Rene Scholl from Scott Aromatics who let me use his GC-MS to analyse the oil.

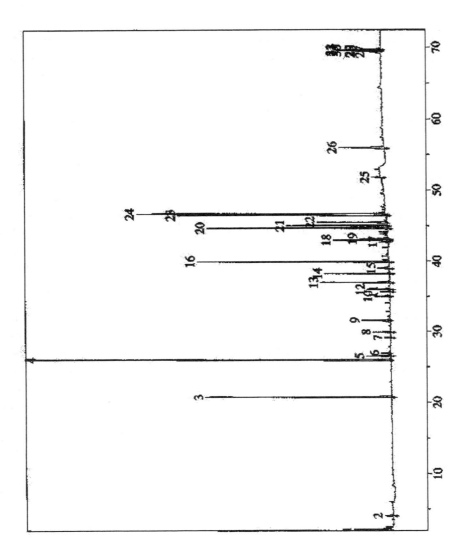

The numbered axis is in minutes, calculated from the time of injection of the sample into the GC machine. When the peaks are all on top of each other, as for peaks 26-33, it is very difficult to distinguish the chemicals from each other, even with mass spectrometry. It is possible to vary the peak separation by varying the speed with which the GC oven heats up. The more quickly it heats up, the more quickly the constituents come out, but if you leave them in there too long, you risk formation of artefacts (chemical changes to the original molecules in the sample).

Peak number	Constituent	Percentage
1	solvent	0.75
2	?ethanol	0.29
3	linalool	7.26
4	benzyl acetate	19.15
5	?	0.9
6	citronellol	0.48
7	geraniol	0.39
8	benzyl alcohol	0.91
9	cis-jasmone	1.25
10	p-cresol	0.65
11	eicosane (a wax)	0.39
12	3-hexenol-1 benzoate	0.93
13	eugenol	3.03
14	methyl palmitate (fatty acid)	2.54
15	jasmine lactone ?	0.45
16	isophytol (a sesquiterpenol)	7.73
17	methyl stearate (fatty acid)	0.29
18	indole	2.83
19	? octadecanoic methyl ester	1.15
20	phytol (a sesquiterpenol)	8.49
21	nerolidol/farnesol	3.58
22	methyl linolenate	2.58
23	phytol	9.35
24	benzyl benzoate	13.25
25	palmitic acid (fatty acid)	0.3
26 -33	? nerolidol and other sesquiterpenols	7.49

MAJOR CONSTITUENTS OF ESSENTIAL OILS

In order to try and present a ready reference table, the following is a list of the three major constituents of each oil which have already appeared at the end of each section in Chapters Three and Four. "Major" means a high percentage of the constituent is present in the oil. I have attempted to include some of the minor components which may have a significant therapeutic or hazardous effect, but this is not the publication for a comprehensive report.

The information here comes from entries in the Boelens BACIS ESO database (1994), and where possible I quote the original article as per that database at the end of the book. The percentages have been rounded up, and if I have found two variations or chemotypes of the oil, I have included both (e.g. Frankincense and Basil).

Name of oil	1	2	3	Distinctive minor constituents
Ajowan *Trachyspermum copticum* L. Link[1]	thymol 61%	para-cymene 15.5%	gamma-terpinene 11%	beta-pinene 3.3%, terpinen-4-ol 1.1%, carvacrol 0.6%, alpha-pinene 0.3%
Angelica (root) *Angelica archangelica*[2]	alpha-pinene 25%	1,8-cineole 14.5%	alpha-phellandrene 13.5%	borneol 1%, ambrettolide 0.3%, pentadecanolide 0.2%, coumarins 2% (Pénoël)
Anise (Spain) *Pimpinella anisum*[3]	(E)-anethole 96%	limonene 0.6%	anisaldehyde 0.56%	anisalcohol 0.4%, (Z)-anethole 0.36%, methyl chavicol 0.3%, linalool 0.26%, coumarins 1% (Pénoël)
Annual Wormwood (Yugoslavia) *Artemisia alba*[4]	artemisia ketone 44.8%	1,8-cineole 9.6%	camphor 6.3%	alpha-ylangene 6.2%, verbenone 2.6%, artemisia alcohol 1.8%, methyl chamazulene 0.3%, benzyl isovalerate 0.3%, ethyl 2-methyl butyrate 0.2%, arteannuin B 0.2%,
Artemisia alba (Belgium)[5]	isopinocamphone 34.6%	camphor 21.1%	1,8-cineole 5.7%	myrtenol 3.8%, T-cadinol 3.4%, myrtenal 3%, pinocamphone 1.5%
Basil (Comoro Islands) *Ocimum basilicum*[6]	methyl chavicol 85%	1,8-cineole 3.25%	para-cymene 2.7%	methyl eugenol 1.3%, linalool 0.96%, eugenol 0.45%, p-methoxycinnamaldehyde 0.4%, spathulenol 0.12%
Basil (Portugal) *Ocimum basilicum*[7]	linalool 38.2%	methyl chavicol 16.4%	beta-caryophyllene 7%	eugenol 5.1%, methyl (E)-cinnamate 4.7%, alpha-terpinyl acetate 4.5%, 1,8-cineole 3.5%, terpinen-4-ol 2.5%, gamma-cadinene 2.28%
Bay (West Indian) *Pimenta racemosa* Mill. JS Moore[8]	eugenol 56%	chavicol 21.6%	myrcene 13%	linalool 1.7%, methyl cinnamate 1.1%, 1-octen-3-ol 1%, 3-octanol 0.6%, terpinen-4-ol 0.3%, 1,8-cineole 0.2%
Bergamot (Calabria) *Citrus bergamia*[9]	limonene 38.4%	linalyl acetate 28%	linalool 8%	gamma-terpinene 8%, beta-pinene 7%, alpha-pinene 1.37%, bergamotene 0.31%, geranyl acetate 0.37%
Black Pepper (Sarawak) *Piper nigrum*[10]	beta-caryophyllene 34.6%	delta-3-carene 16%	limonene 14.5%	beta-pinene 7.7%, alpha-pinene 3.7%, alpha-copaene 2.94%, delta-elemene 1.8%, cadinol 0.9%, alpha-cubebene 0.2%, torreyol 0.18%

111

Name of oil	1	2	3	Distinctive minor constituents
Buchu *Barosma betulina* Bertl. (also *Agathosma betulina*)[11]	(-)-menthone 35%	diosphenol 12%	(-)-pulegone 11%	limonene 10%, piperitone oxide 9.5%, (+)-menthone 9%, (+)-isopulegone 3%, 8-mercapto-p-3-menthanone 1%
Cajeput *Melaleuca leucadendron* L.[12]	1,8-cineole 41.1%	alpha-terpineol 8.7%	para-cymene 6.8%	terpinolene 4.6%, gamma-terpinene 4.6%, limonene 4.1%, linalool 3.6%, eudesmols 1.9%, terpinen-4-ol 1.5%, guaiol 1.2%, beta-pinene 0.8%
Camphor (Japan) *Cinnamomum camphora* L. Nees et Eberm.[13] (probably Yellow or Brown camphor)	camphor 51.5%	safrole 13.4%	1,8-cineole 4.75%	piperitone 2.4%, beta-caryophyllene 1.4%, cinnamyl alcohol 0.18%, furfural 0.16%, eugenol 0.12%, cinnamaldehyde 0.08%
Caraway (Netherlands) *Carum carvi* L.[14]	(+)-carvone 50%	limonene 46%	cis-dihydrocarvone 0.48%	myrcene 0.39%, carveol 0.38%, perillaldehyde 0.14%,
Cardamom (Reunion) *Cardamomum elettaria* L. Maton var. *alpha-minor*[15]	1,8-cineole 48.4%	alpha-terpinyl acetate 24%	limonene 6%	alpha-terpineol 1.9%, geraniol 1.6%, terpinen-4-ol 1.4%, linalool 1.2%, borneol 0.3%, geranial 0.3%, linalyl acetate 0.2%, carvone 0.15%, neral 0.15%
Catmint *Nepeta cataria*[16]	nepetalactone 1 80.5%	nepetalactone 2 10%	beta-caryophyllene 4.4%	alpha-humulene 0.75%, caryophyllene oxide 0.6%, beta-farnesene 0.2%
Cedarleaf (Canada) *Thuja occidentalis*[17]	alpha-thujone 56%	fenchone 15%	beta-thujone 14.7%	bornyl acetate 4.8%, camphor 2.9%, terpinen-4-ol 2.4%, para-cymene 0.9%
Cedarwood Texas *Juniperus mexicana*[18]	thujopsene 32%	alpha-cedrene 24.1%	cedrol 16%	beta-cedrene 5.6%, gamma-eudesmol 2.45%, beta-caryophyllene 0.5%
Cedarwood Virginia *Juniperus virginiana*[19]	cedrol 26%	alpha-cedrene 24.5%	thujopsene 15%	beta-cedrene 6.7%, gamma-eudesmol 5.2%, beta-caryophyllene 2.6%, alpha-pinene 0.4%
Cinnamon Bark *Cinnamomum zeylanicum* Blume[20]	cinnamaldehyde 74%	eugenol 8.8%	cinnamyl acetate 5.1%	linalool 2.3%, 1,8-cineole 1.65%, benzyl benzoate 1%, alpha-terpineol 0.4%

Name of oil	1	2	3	Distinctive minor constituents
Cinnamon Leaf *Cinnamomum zeylanicum* Blume[21]	eugenol 87%	benzyl benzoate 2.6%	beta-caryophyllene 1.8%	linalool 1.5%, cinnamaldehyde 1.3%, eugenyl acetate 1%, cinnamyl acetate 0.8%, safrole 0.65%, cinnamyl alcohol 0.6%
Citronella (Ceylon) *Cymbopogon nardus* L. Rendle[22]	geraniol 18%	limonene 9.7%	citronellol 8.4%	methyl isoeugenol 7.2%, borneol 6.6%, citronellal 5.2%, geranyl formate 4.2%, citronellyl acetate 1.9%, methyl eugenol 1.7%, geranyl butyrate 1.5%
Citronella (Java) *Cymbopogon winterianus* Jowitt[23]	citronellal 36.8%	geraniol 21.4%	citronellol 15%	citronellyl acetate 2.5%, elemol 2.1%, isopulegol 2.1%, geranial 0.8%, neral 0.6%, eugenol 0.6%, alpha-bergamotene 0.1%
Clary Sage (USA) *Salvia sclarea*[24]	linalyl acetate 49%	linalool 24%	germacrene D 3%	alpha-terpineol 3%, geranyl acetate 2.5%, geraniol 2.4%, sclareol 0.31%, 3-hexenol 0.2%, spathulenol 0.03%,
Clove Bud (Madagascar) *Eugenia caryophyllus* (Spreng.) Bullock[25]	eugenol 76.6%	beta-caryophyllene 9.8%	eugenyl acetate 7.6%	alpha-humulene 1.2%, isoeugenol 0.19%, alpha-terpinyl acetate 0.17%, alpha-copaene 0.15%, alpha-cubebene 0.15%, methyl benzoate 0.09%
Dwarf Pine *Pinus mugo ssp. pumilia*[26]	delta-3-carene 35%	beta-phellandrene 15%	alpha-pinene 13.1%	beta-caryophyllene 3.32%, bornyl acetate 1.25%, p-cymene 0.72%, germacrene D 0.5%, T-cadinol 0.01%
Elemi *Canarium luzonicum*[27]	limonene 54%	alpha-phellandrene 15.1%	elemol 15%	elemicin 3.5%, 1,8-cineole 2.5%, myrcene 2.4%, methyl eugenol 0.3%, carvone 0.2%
Eucalyptus *Eucalyptus globulus* (Spain)[28]	1,8-cineole 66.1%	alpha-pinene 14.7%	limonene 3%	aromadendrene 2.2%, pinocarvone 1%, alpha-terpinyl acetate 0.88%, globulol 0.47%, alpha-terpineol 0.44%, beta-pinene 0.35%, terpinen-4-ol 0.13%, 3-methylbutanal 0.15%, isoamyl isovalerate 0.02%
Fir balsam (Canada) *Abies balsamea*[29]	beta-pinene 30%	delta-3-carene 21.45%	bornyl acetate 11.85%	limonene 11%, borneol 0.35%, piperitone 0.15%, terpinen-4-ol 0.1%

Name of oil	1	2	3	Distinctive minor constituents
Frankincense (Somalia) *Boswellia carterii*[30]	alpha-pinene 34.53%	alpha-phellandrene 14.6%	para-cymene 14%	1,8-cineole 1%, beta-bourbenene 0.32%, beta-elemene 0.1%, beta-caryophyllene 0.08%, allo-aromadendrene 0.03%
Frankincense (Somalia) *Boswellia carterii*[31]	octyl acetate 60%	1-octanol 12.7%	alpha-pinene 3.5%	incensol 2.7%, linalool 2%, isocembrene 1.8%, 1,8-cineole 1.6%, bornyl acetate 1.1%, decyl acetate 0.3%
Geranium Bourbon *Pelargonium graveolens*[32]	citronellol 21.2%	geraniol 17.45%	linalool 12.9%	citronellyl formate 8.37%, geranyl formate 7.55%, isomenthone 7.2%, guaia-6,9-diene 3.9%, menthone 1.5%, geranyl butyrate 1.3%, citronellyl butyrate 1.26%, cis-rose oxide 0.64%, furopelargone A 0.37%, 2-phenylethyl tiglate 0.43%, trans-rose oxide 0.21%
German Chamomile (Bulgaria) *Chamomilla recutita* L. Rauschert (also *Matricaria chamomilla*)[33]	farnesene 27%	chamazulene 17%	alpha-bisabolol oxide B 11%	alpha-bisabolol 9.5%, alpha-bisabolol oxide A 8%, delta-cadinene 5.2%, alpha-muurolene 3.4%, unknown dicycloether 0.7%
Ginger (China) *Zingiber officinale* Roscoe[34]	ar-curcumene 16.3%	alpha-zingiberene 14.2%	beta-sesquiphellandrene 10.6%	bisabolenes 10.7%, citral 4%, geraniol 1.7%, zingiberenol 1.4%, citronellol 1.2%, beta-eudesmol 0.9%, 2-undecanone 0.3%, 6-methyl-5-hepten-2-one 0.2%, hexanal 0.05%, decanal 0.04%
Grapefruit (Israel) *Citrus paradisi*[35]	limonene 93%	myrcene 1.97%	alpha-pinene 0.59%	nootkatone 0.3%, octanal 0.29%, decanal 0.27%
Ho leaf *Cinnamomum camphora* Sieb. ssp.*formosana* var. *brientalis*[36]	linalool 95%	camphor 0.4%	limonene 0.2%	linalyl acetate 0.2%, 1,8-cineole 0.1%, alpha-terpineol 0.05%

Name of oil	1	2	3	Distinctive minor constituents
Hyssop (France) *Hyssopus officinalis*[37]	isopinocamphone 32.6%	beta-pinene 22.9%	pinocamphone 12.2%	methyl myrtenate 4.8%, myrtenol methyl ether 2.7%, spathulenol 2.2%, elemol 1.7%, methyl chavicol 1.3%, allo-aromadendrene 0.5%, methyl eugenol 0.5%
Imortelle/Everlasting *Helichrysum italicum* G. Don.[38]	alpha-pinene 21.7%	gamma-curcumene 10.4%	italidiones 8%	neryl acetate 6.1%, beta-selinene 6%, beta-caryophyllene 5%, alpha-curcumene 4%, italicene 4%, alpha selinene3.6%, bisabolane hydroxyketones 2%, isoitalicene 1.55, neryl propionate 1.2%, 2-methylbutyl angelate 0.6%, nerolidol 0.3%, borneol 0.2%, italicene epoxide 0.2%, italicene ethers 0.2%
Jasmine (France) *Jasminum grandiflorum*[39]	benzyl acetate 22%	benzyl benzoate 14.5%	phytyl acetate 10.2%	linalool 6.4% methyl cis-jasmonate 3%, (Z)-3-hexenyl benzoate 2.6%, cis-jasmone 2.6%, indole 2.4%, methyl anthranilate 2%, benzyl alcohol 1.56%
Jasmine (Malati) *Jasminum sambac* L. Aiton[40]	linalool 25%	(-)-germacra-1,6-dien-5-ol 20%	alpha-farnesene 12.5%	phytol 8%, (Z)-3-hexenyl benzoate 5%, benzyl acetate 3%, benzyl alcohol 3%, methyl anthranilate 3%, benzyl benzoate 0.5%, indole 0.1%, jasmone 0.1%, p-cresol 0.05%, eugenol 0.05% methyl jasmonate 0.01%
Juniper berry *Juniperus communis*[41]	alpha-pinene 33%	myrcene 11%	beta-farnesene 10.5%	gamma-elemene 2.9%, beta-caryophyllene 2.7%, beta-pinene 2.5%, sabinene hydrate 0.9%, aromadendrene 0.6%, bornyl acetate 0.4%, verbenone 0.2%
Lavandin (Abrial quality, France) *Lavandula hybrida* Rev. (L. x intermedia Emeric ex Loiseleur)[42]	linalool 33.5%	linalyl acetate 27.1%	camphor 9.5%	1,8-cineole 8.1%, borneol 2.5%, beta-caryophyllene 2.4%, lavandulyl acetate 1.7%, lavandulol 0.9%, 1-octenyl-3-acetate 0.5%, 3-octanone 0.4%, 1-octen-3-ol 0.3%, hexyl butyrate 0.3%

Name of oil	1	2	3	Distinctive minor constituents
Lavender (France) *Lavandula angustifolia* Mill.[43]	linalyl acetate 40%	linalool 31.5%	(Z)-beta-ocimene 6.7%	beta-caryophyllene 5.16%, lavandulyl acetate 4.2%, terpinen-4-ol 4%, 3-octanone 1.5%, lavandulol 0.7%, 1,8-cineole 0.69%, camphor 0.3%
Lemon (Argentina) *Citrus limonum*[44]	limonene 70%	beta-pinene 11%	gamma-terpinene 8%	citral 1.59%, trans-alpha-bergamotene 0.4%, geranyl acetate 0.22%, nonanal 0.12%
Lemon Balm/Melissa *Melissa officinalis* L.[45]	geranial 45%	neral 35%	6-methyl-5-hepten-2-one 3%	beta-caryophyllene 2%, citronellal 2%, geranyl acetate 2%, aesculetine 0.5% (Pénoël), damascone 0.05%, eugenol 0.05%, vetispirane 0.05%
Lime (Persian) *Citrus latifolia* Tanaka[46]	limonene 58%	gamma-terpinene 16%	beta-pinene 6%	alpha-terpineol 2.15%, p-cymene 1.56%, 1,8-cineole 1%, terpinen-4-ol 0.46%, geranial 0.12%
Mandarin (Italy) *Citrus deliciosa* Tenore[47]	limonene 71%	gamma-terpinene 18.54%	alpha-pinene 2.39%	alpha-sinensal 0.2%, octanal 0.17%, methyl N-methyl anthranilate 0.15%, decanal 0.07%, nonanal 0.07%
May Chang *Litsea cubeba* (berry)[48]	geranial 40%	neral 33.8%	limonene 8.3%	6-methyl-5-hepten-2-one 4.4%, linalool 1.7%, linalyl acetate 1.6%, geraniol 1.5%, alpha-terpinyl acetate 0.16%,
Mugwort (Germany) *Artemisia absinthium*[49]	beta-thujone 46%	sabinyl acetate 25%	trans-sabinol 3.2%	lavandulyl acetate 2.7%, alpha-thujone 2.7%, geranyl propionate 1.43%, 1,8-cineole 0.7%, lavandulol 0.5%, terpinen-4-ol 0.4%, gamma-cadinene 0.3%
Myrrh gum (headspace) *Commiphora myrrha* (Nees) Engler[50]	delta-elemene 28.7%	alpha-copaene 10%	beta-elemene 6.1%	methyl isobutyl ketone 5.6%, 2-methyl-5-isopropenyl furan 4.6%, 3-methyl-2-butenal 2.2%, 2-methylfuran 1.9%, 4,4-dimethyl-2-butenolide 1%, dihydrocurzerenone 1.1%, assorted furans, furfurals etc.

Name of oil	1	2	3	Distinctive minor constituents
Myrtle (Spain, wild) Myrtus communis L. [51]	myrtenyl acetate 35.9%	1,8-cineole 29.9%	alpha-pinene 8.1%	limonene 7.5%, alpha-terpineol 4.1%, methyl eugenol 2.3%, carvacrol 0.6%, myrtenol 0.58%, linalyl acetate 0.53%, isobutyl isobutyrate 0.4%,
Neroli bigarade Citrus aurantium var. amara [52]	linalool 37.5%	limonene 16.6%	beta-pinene 11.8%	geraniol 4.25%, linalyl acetate 2.8%, nerolidol 2.6%, geranyl acetate 1.7%, neryl acetate farnesol 1%, 0.9%, terpinen-4-ol 0.75%, alpha-terpinyl acetate 0.2%, indole 0.1%, methyl N-methyl anthranilate 0.1%, cis-jasmone 0.05%
Niaouli (Madagascar) Melaleuca quinquinervia Cav. [53]	1,8-cineole 41.8%	viridiflorol 18.1%	limonene 5.5%	beta-caryophyllene 5%, alpha-pinene 5%, alpha-terpineol 5%, ledol 2.5%, caryophyllene oxide 0.6%, (E)-nerolidol 0.4%
Nutmeg Myristica fragrans [54]	alpha-pinene 22%	sabinene 18.55%	beta-pinene 15.55%	terpinen-4-ol 7.85%, myristicin 6%, gamma-terpinene 5.1%, safrole 2%, alpha-terpineol 1%, eugenol 0.2%, isoeugenol 0.2%
Orange Sweet (Brazil) Citrus sinensis [55]	limonene 89%	myrcene 1.71%	beta-bisabolene 1.29%	1-nonanol 0.67%, linalool 0.35%, neral 0.25%, decanal 0.2%, geranial 0.19%, linalyl acetate 0.14% auraptene 0.09% (Pénoël)
Oregano (Greek) Origanum vulgare L. ssp. viride (Boiss.) [56]	thymol 85.6%	carvacrol 4.3%	gamma-terpinene 2.7%	beta-caryophyllene 2.3%, para-cymene 2.3%, camphor 0.1%, 1,8-cineole 0.1%
Palmarosa (India) Cymbopogon martinii Stapf. var. motia [57]	geraniol 80%	geranyl acetate 8.25%	linalool 2.79%	beta-caryophyllene 1.76%, farnesol 1%, neral 0.4%, alpha-farnesene 0.25%, gamma-selinene 0.24%, geranyl butyrate 0.15%
Patchouli (Indonesia) Pogostemon cablin Benth. [58]	patchouli alcohol 33%	alpha-patchoulene 22%	beta-caryophyllene 20%	beta-patchoulene 13%, beta-elemene 6%, norpatchoulenol 1%, pogostol 0.4%, caryophyllene oxide 0.3%, delta-guaiene 0.3%, patchoulenone 0.05%, patchouli oxide 0.05%, patchouli pyridine 0.05%

Name of oil	1	2	3	Distinctive minor constituents
Pennyroyal *Mentha pulegium*[59]	(+)-pulegone 63.5%	(+)-isomenthone 19.7%	(+)-neoisomenthol 5.7%	3-octanol 2%, (-)-isopulegone 0.7%, limonene 0.7%, piperitenone 0.6%, 3-octyl actate 0.12%, menthofuran 0.01%
Peppermint (USA) *Mentha piperita* L. var. Mitcham[60]	menthol 42.8%	menthone 19.4%	sabinene hydrate 6.6%	1,8-cineole 5.2%, neomenthol 4.2%, isomenthone 3.2% beta-caryophyllene 2.3%, menthofuran 2%, limonene 1.6%, pulegone 0.9%
Petitgrain bigarade *Citrus aurantium* L. ssp. *amara*[61]	linalyl acetate 45.5%	linalool 24.1%	alpha-terpineol 5.2%	geranyl acetate 4.2%, limonene 4%, neryl acetate 2.2%, geraniol 1,8% beta-caryophyllene 1.6%, nerol 8%, 2-phenylethanol 0.2%, methyl N-methyl anthranilate 0.1%, indole 0.05%, 2-methoxy-3-isobutyl pyrazine 0.01%
Pine *Pinus pinaster*[62]	alpha-pinene 44.1%	beta-pinene 29.5%	myrcene 4.69%	beta-caryophyllene 3.45%, delta-3-carene 3.34%, alpha-terpineol 1.35%, alpha-humulene 0.54%
Pine *Pinus sylvestris*[63]	alpha-pinene 42%	delta-3-carene 20.5%	limonene 5.2%	cadinene 4.76%, germacra-1-(10)-E,5E-dien-4-ol, 1.89%, T-cadinol 0.56%, bornyl acetate 0.12%
Roman Chamomile (Japan) *Anthemis nobilis*[64]	isobutyl angelate 35.9%	2-methylbutyl angelate 15.3%	methallyl angelate 8.7%	isobutyl isobutyrate 4.9%, isoamyl angelate 4.34%, pinocarvone 3.59%
Rose (Bulgaria) *Rosa damascena* Mill. (otto)[65]	citronellol 33.4%	stearopten waxes 24%	geraniol 18%	nerol 5.9%, linalool 2.1%, 2-phenyl ethyl alcohol 1.3%, eugenol 1.5%, farnesol 0.87%, methyl eugenol 0.5%, geranial 0.5%, cis-rose oxide 0.4%, 3-hexenal 0.26%, carvone 0.22%, trans-rose oxide 0.4%, beta-damascenone 0.05%
Rose (Egypt) *Rosa damascena* Mill.[66]	2-phenyl ethyl alcohol 37.9%	geraniol 15.8%	citronellol 12.6%	farnesol 6.3%, nerol 4%, linalool 2.2%, eugenol 1.2%, alpha-ionone 1%

Name of oil	1	2	3	Distinctive minor constituents
Rosemary (Spain)[67] (camphor) *Rosmarinus officinalis*	alpha-pinene 22%	camphor 17%	1,8-cineole 17%	verbenone 4%, borneol 2%, bornyl acetate 1.5%, terpinen-4-ol 1.5%, alpha-terpineol 1.5%
Rosemary (Tunisia)[68] (cineole) *Rosmarinus officinalis*	1,8-cineole 51.3%	camphor 10.6%	alpha-pinene 10%	borneol 7.7%, alpha-terpineol 3.9%, terpinen-4-ol 1%, bornyl acetate 0.8%, alpha-humulene 0.7%, verbenone 0.05%
Rosewood *Aniba rosaeodora*[69]	linalool 85.3%	alpha-terpineol 3.5%	cis-linalool oxide 1.5%	trans-linalool oxide 1.3%, 1,8-cineole 1%, geranyl acetate 0.14%
Rue (Egypt) *Ruta graveolens*[70]	2-undecanone 49.2%	2-nonanone 24.7%	2-nonyl acetate 6.2%	bergapten 7%, limonene 6%, 2-decanone 2.8%, pregeijerene 2.1%, 2-dodecanone 1.1%
Sage (Dalmatia) *Salvia officinalis*[71]	alpha-thujone 37.1%	beta-thujone 14.2%	camphor 12.3%	1,8-cineole 12%, alpha-pinene 3.9%, alpha-humulene 3.8%, bornyl acetate 0.86%, terpinen-4-ol 0.2%
Sandalwood (India)[72] *Santalum album*	cis-alpha-santalol 50%	cis-beta-santalol 20.9%	epi-beta-santalol 4.1%	alpha-santalal 2.9%, cis-lanceol 1.7%, trans-beta-santalol 1.5%, spirosantalol 1.2%, santalenes 1.7%, cis-nuciferol 1.1%, beta-santalal 0.56%, eka-santalals 0.08%
Savory, summer (Italy)[73] *Satureja hortensis* L	carvacrol 48%	gamma-terpinene 28%	para-cymene 7%	alpha-terpinene 2.8%, beta-caryophyllene 1.25%, alpha-pinene 1.2%, beta-bisabolene 0.65%, methyl chavicol 0.15%
Spearmint (Greece)[74] *Mentha spicata*	(-)-carvone 42.8%	dihydrocarvone 15.7%	1,8-cineole 5.8%	perillyl alcohol 4.5%, alpha-terpinenyl acetate 4.5%, beta-bourbenene 2.6%, beta-caryophyllene 2.5%
Spike Lavender (Spain) *Lavandula latifolia* Medicus[75]	1,8-cineole 36.3%	linalool 30.3%	camphor 8%	borneol 2.8%, alpha-terpineol 2.6%, caryophyllene oxide 2.4%, coumarin 2.4%, linalool oxide 0.5%, isoborneol 0.3%
Star Anise (China) *Illicium verum* Hook. F.[76]	(E)-anethole 71.5%	foeniculin 14.5% (an azulene)	methyl chavicol 5%	limonene 1.4%, linalool 0.6%, nerolidol 0.5% cinnamyl acetate 0.23%, (Z)-anethole 0.01%

Name of oil	1	2	3	Distinctive minor constituents
Sweet Fennel (Turkey) *Foeniculum vulgare* Mill. var. *dulce*[77]	(E)-anethole 80%	limonene 6%	methyl chavicol 4.5%	fenchone 2%, anisaldehyde 1%, 1,8-cineole 0.4%, (Z)-anethole 0.3%, carvone 0.1%, citral 0.1%, octanal 0.1%
Sweet Marjoram *Marjorana hortensis* Moench (*Origanum majorana*)[78]	terpinen-4-ol 36.3%	cis-sabinene hydrate 15.9%	para-cymene 9.5%	alpha-terpineol 8.2%, linalool 3.9%, linalyl acetate 3.5%, bicyclogermacrene 2.5%, beta-caryophyllene 2%
Tansy (Belgium) *Tanacetum vulgare*[79]	beta-thujone 50%	trans-chrysanthemyl acetate 20%	camphor 6.4%	germacrene D 5%, alpha-thujone 1.8%, 1,8-cineole 0.5%, alpha-pinene 0.5% eugenol 0.4%,
Tarragon (USA) *Artemisia dranunculus* (French)[80]	methyl chavicol 80%	beta-ocimenes 14%	limonene 2.5%	alpha-pinene 0.5%, methyl eugenol 0.48%, elemicin eugenol 0.2%, 0.05%
Tea Tree *Melaleuca alternifolia*[81]	terpinen-4-ol 45.4%	gamma-terpinene 15.7%	alpha-terpinene 7.1%	para-cymene 6.2%, alpha-terpineol 5.3%, 1,8-cineole 3%, alpha-pinene 2.1%, limonene 1.4%
Thyme (Italy) *Thymus vulgaris*[82]	thymol 27.4%	para-cymene 21.9%	gamma-terpinene 12%	beta-caryophyllene 3.3%, linalool 2.6%, 1,8-cineole 2.2%, carvacrol 1.1%, alpha-thujone 0.25%, delta-cadinene 0.18%
Tumeric (Indonesia) *Curcuma longa*[83]	turmerone 29.5%	ar-turmerone 24.7%	turmerol 20%	beta-curcumene 2.5%, alpha-atlantone 2.4%, curcuphenol 0.6%, beta-bisabolol 0.3%
Vanilla *Vanilla fragrans* Ames[84]	vanillin 85% (ether)	4-hydroxybenzaldehyde 8.5%	4-hydroxybenzyl methyl ether 1%	not specified, but incl. alkyl benzenes and esters
Vetiver *Vetivera zizanoides* Stapf.[85]	vetiverol 50%	vetivenes 20%	alpha-vetivol 10%	vetivones 10%, khusimol 1%, khusimone 1%, vetiselinene 1%, vetispirenes 2%, vetiazulene 0.1%
Wintergreen (China) *Gaultheria procumbens*[86]	methyl salicylate 90%	safrole 5%	linalool 2%	1,8-cineole 1%, camphor 0.5%, alpha-pinene 0.5%, eugenol 0.2% ethyl salicylate 0.1%

Name of oil	1	2	3	Distinctive minor constituents
Yarrow *Achillea millefolium*[87]	camphor 17.7%	sabinene 12.3%	1,8-cineole 9.5%	alpha-pinene 9.4%, iso-artemisia ketone 8.6%, beta-pinene 7.3%, terpinen-4-ol 4.3%, borneol 2.5%,
Ylang ylang *Cananga odorata*[88]	linalool 19%	beta-caryophyllene 10.5%	germacrene D 10.2%	p-cresyl methyl ether 8.7%, benzyl benzoate 7.3%, geranyl acetate 6.7%, benzyl acetate 4.6%, benzyl salicylate 2%, farnesol 1.8%, cadinols 1.8%, eugenol 0.3%, 3-methyl-2-butenyl acetate 0.13%,

CHAPTER SEVEN

REPRESENTING ESSENTIAL OIL CONSTITUENTS

The exact structures of compounds are usually deduced by a combination of anaylsis techniques including infra-red spectroscopy, which determine the number of single, double and triple carbon bonds, the presence of any aromatic rings, and the presence of different oxygenated functional groups.

The common names of compounds, such as "ocimene" are more user-friendly for aromatherapists, but leaving it at that can blur the differences between molecules which are very similar in structure (i.e. are isomers of each other). From a therapeutic point of view, it becomes useful to know which isomer of a compound is present is when one isomer is biologically active, and the other is not. For most essential oil constituents, this is an unexamined area of research.

Isomers

The differences between isomers can be in the positioning of a double bond, whereupon you get the distinction of alpha-, beta-, gamma- or delta-attached to the front of the name. These prefixes are sometimes written as the Greek letters instead. An example is alpha-pinene and beta-pinene found in Pine oil (see end of this chapter, section Monoterpenes).

When you have a carbon double bond, there are several ways that the atoms adjoining the carbons can be arranged. If the longer chains come off the same side of the double bond, the molecule can be called a cis-isomer. If they come off opposite sides, the molecule will be called a trans-isomer[99]. The simplified example below demonstrates it using letter A to denote shorter chains, and letter B to denote longer chains.

cis-isomer   trans-isomer

---

[99] Another way of denoting cis- and trans- is (Z) - and (E)- which stand for the German words "zusammen" which means "together" and "entgegen" which means "opposite".

If the two molecules are identical in two dimensional notation, but when you get them made up in three dimensional model form you couldn't superimpose them one on the other (this is like the difference between left and right handed gloves), then they are known as stereoisomers, enantiomers or optical isomers. The reason for the last name, "optical" isomers, is that if you pass polarised light through a pure solution of the (+)-isomer, it will twist the light in a clockwise or dextrorotatory fashion. A pure solution of the (-)-isomer will twist the polarised light in an anticlockwise or levorotatory fashion.

Sometimes, as in the case of limonene, you will find written *d*-limonene and *l*-limonene, which are in fact (+)-limonene and (-)-limonene respectively. There is a potential for confusion when looking at the names of amino acids and sugars, because they usually have a capital D or L preceding the name. This refers to what is known as their absolute configuration in relation to the molecule glyceraldehyde, and does not necessarily relate to the optical activity of the molecule with polarised light.

Chirality
When a molecule has two chiral carbon atoms (where there are four different groups coming off each bond), sometimes you will see the use of R- and S-, which stand for *rectus*, meaing "right" and *sinister*, meaning "left". This allows for comparison between two stereoisomers, and the exact representation of their three-dimensional structure.

If you take the smallest group attached to the carbon atom, and have that pointing away from you, then you are left with an almost planar view of the other three attachments. You start with the largest atom (e.g. and oxygen atom is larger than a carbon atom), and if not the largest atom, then the group with the largest atoms closest to the chiral carbon, and work your way around the three attachements.

If you find yourself going clockwise, then that carbon atom is designated R-, and if you go anti-clockwise it is designated S-. An example is (1S,3S,4R)-(+)-menthol and (1R,3R,4S)-(-)-menthol, which are shown below. The two pictures of (-)-menthol have just been turned 180° in space, and in fact both of them are mirror images of the (+)-menthol and cannot be superimposed on it. In these drawings, the bold bonds are coming out of the page towards you and the dashed bonds are receding into the page.

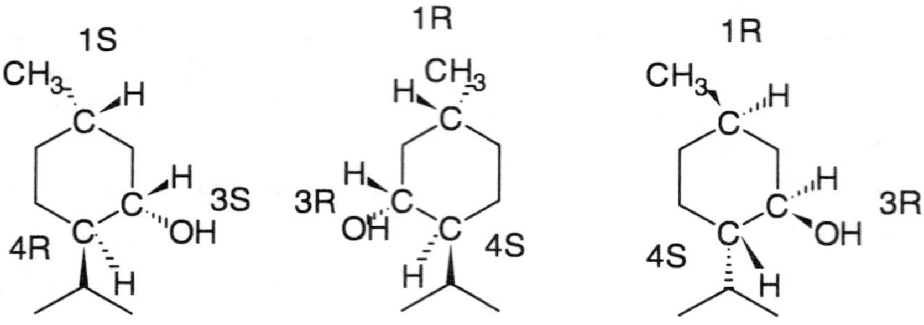

1S

(+)-menthol

1R

(-)-menthol

1R

(-)-menthol

The biological significance of stereoisomers is that in living systems, there seems to be a strong selection of the L-isomers of amino acids. This is caused by the action of special enzymes which make only the L-isomers. In a chemistry lab, if you make a particular amino acid from scratch, you are going to naturally get a mixture of D- and L-isomers.

Such a mixture (known as a racemic mixture) will be less biologically active than the pure L-isomers, as the body cannot use the D-isomers. It remains to be seen whether there is a similar biological effect between optical isomers of essential oils, although already it is well known that different optical isomers can have different smells. For example, (+)-menthol smells minty, musty phenolic and medicinal, whereas (-)-menthol smells fresh, cooling and strongly minty according to Boelens et al. (1993)[100]. They also say that the cooling taste effect is significantly stronger for the (-)-menthol, which may have implications for other therapeutic effects.

Chemical names
An example of the common name of a molecule and its systematic chemical name is alpha-terpinene and 1-methyl-4(1-methylethyl)-1,3-cyclohexadiene. The numbers indicate the location of the methyl and ethyl groups and the double bonds as you go around the closed ring or along the chain. The aim of the numbering it to keep the numbers as low as possible. "Cyclo-" means there is a closed ring, "hexa-" means there

---

[100] Boelens MH, Boelens H, van Gemert LJ, (1993) "Sensory properties of optical isomers" *Perfumer and Flavorist*, v18, pp. 2-16.

are six carbon atoms in the ring, and "-diene" means there are two double bonds at the location indicated by the numbers preceding the "cyclo-".

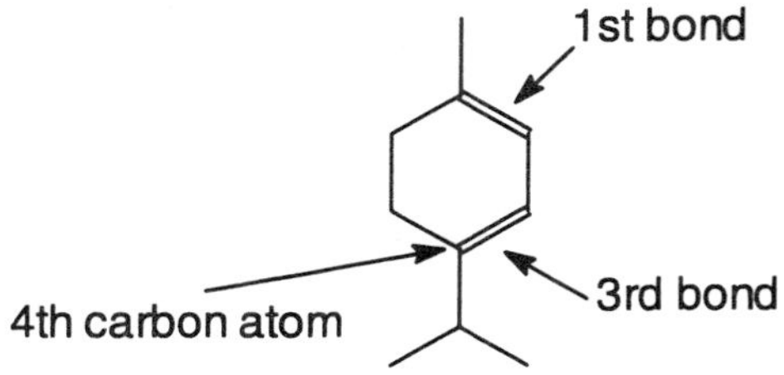

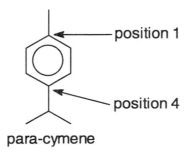

alpha-terpinene

Another term which is usually found in relation to molecules which have benzene rings, is the term "para-", or sometimes just "p-". This refers to the position of two functional groups attached to the benzene ring, in relationship to each other. An example is the molecule para-cymene shown below. A chemist would say the methyl group (position 1) is "para-" to the isopropyl group (position 4). The other positions are known as "meta-" for position 3, and "ortho-" for position 2.

para-cymene

Monoterpenes

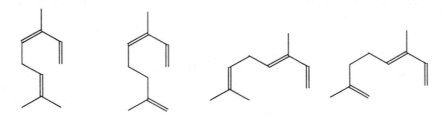

cis-beta-ocimene   cis-alpha-ocimene   trans-beta-ocimene   trans-alpha-ocimene

myrcene   alpha-terpinene   gamma-terpinene   limonene   beta-phellandrene

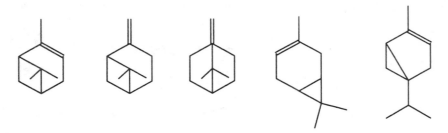

alpha-pinene   beta-pinene   camphene   delta-3-carene   alpha-thujene

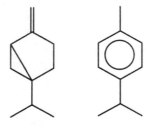

sabinene   para-cymene

Sesquiterpenes

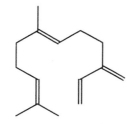

beta-farnesene

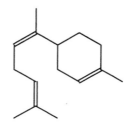

alpha-bisabolene

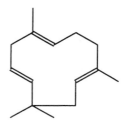

alpha-humulene

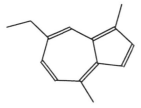

chamazulene

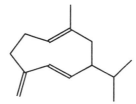

germacrene-D

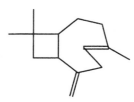

beta-caryophyllene

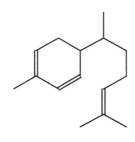

zingiberene

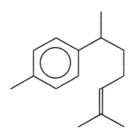

alpha-curcumene

Monoterpenols

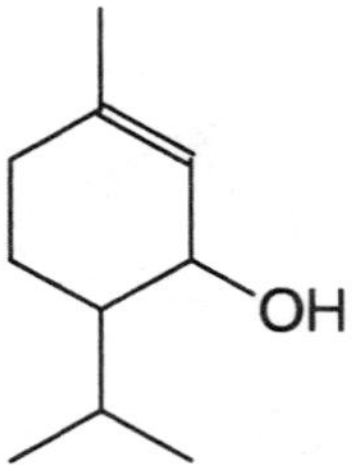

piperitol 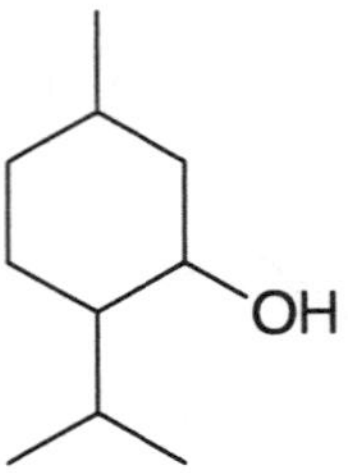   menthol 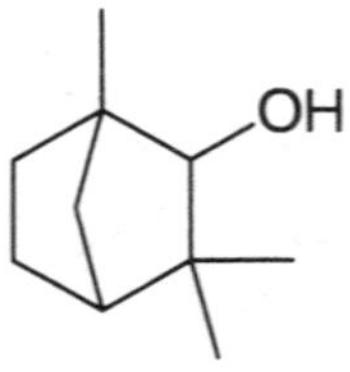   alpha-fenchol 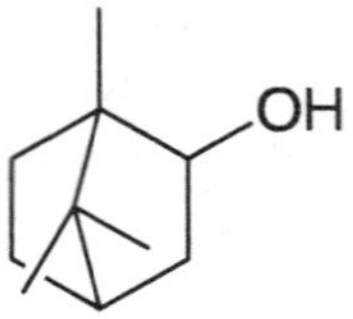   borneol

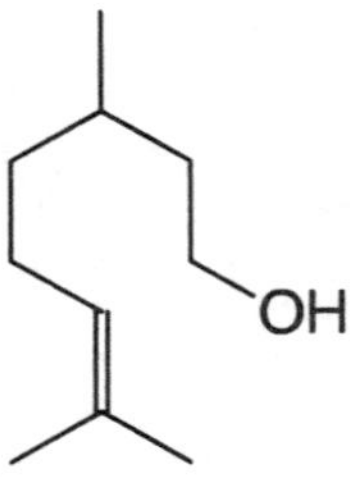

citronellol 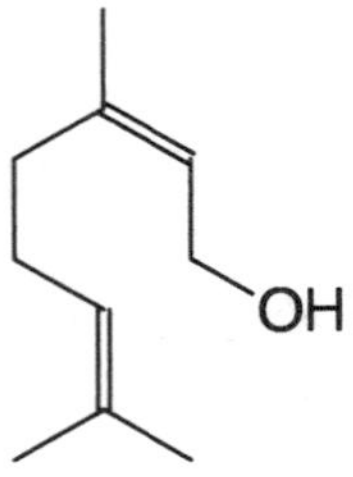   nerol 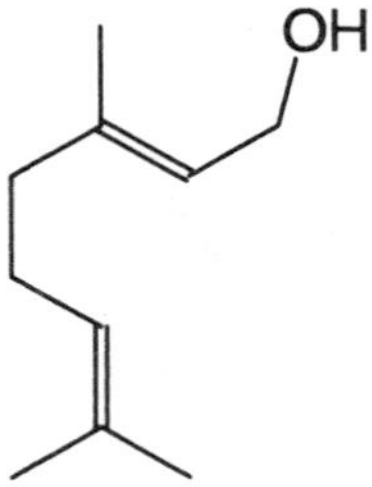   geraniol 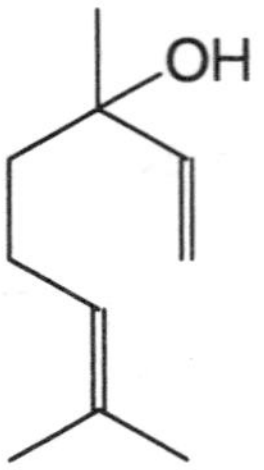   linalool

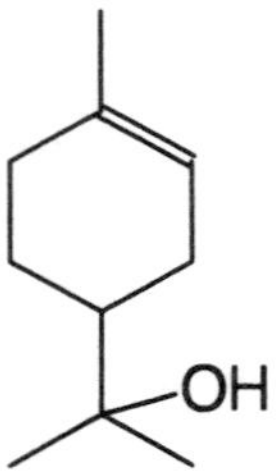

alpha-terpineol 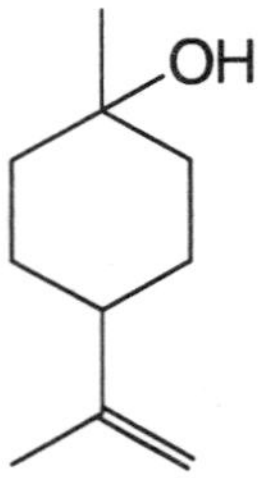   beta-terpineol 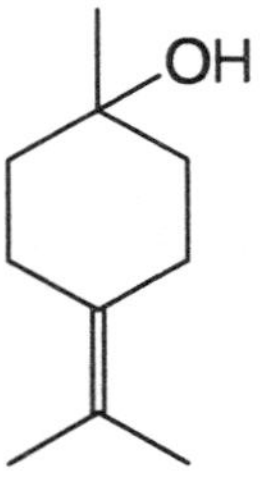   gamma-terpineol 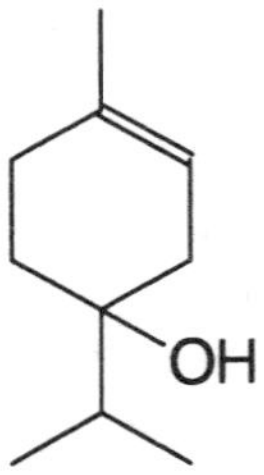   terpinen-4-ol

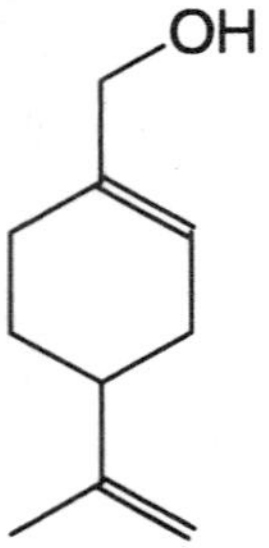

perillyl alcohol 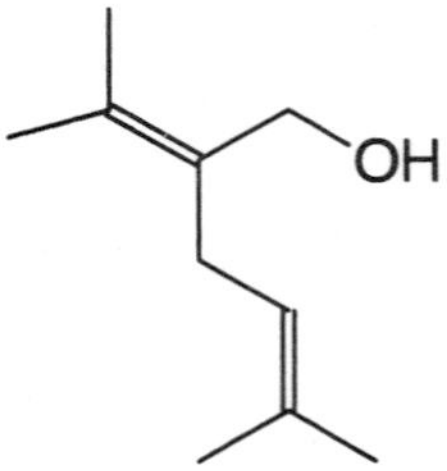   lavandulol 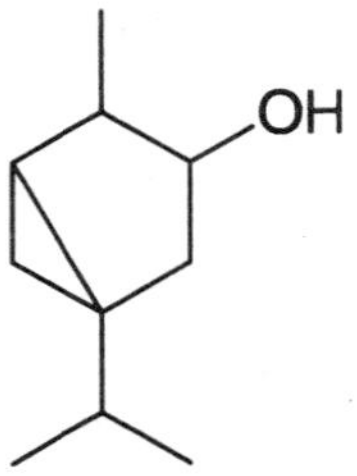   thujanol

Sesquiterpenols

alpha-cadinol   alpha-bisabolol   beta-farnesol

nerolidol   viridflorol   patchoulol

cedrol   widdrol   beta-santalol

Phenols

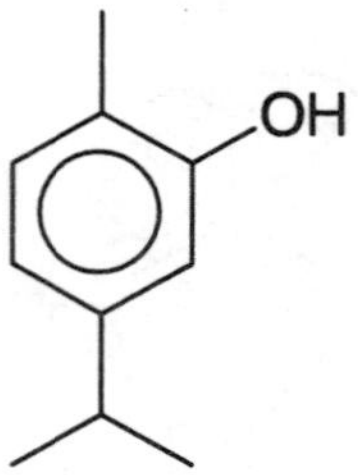

| carvacrol | thymol | chavicol | eugenol |

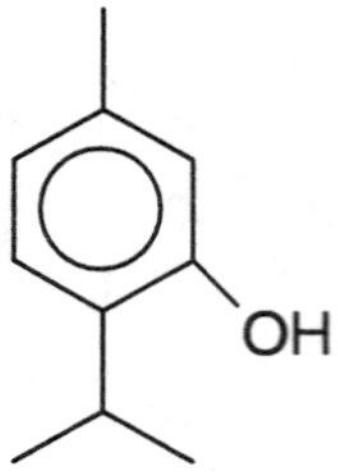

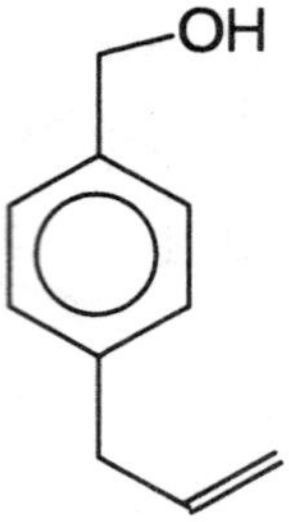

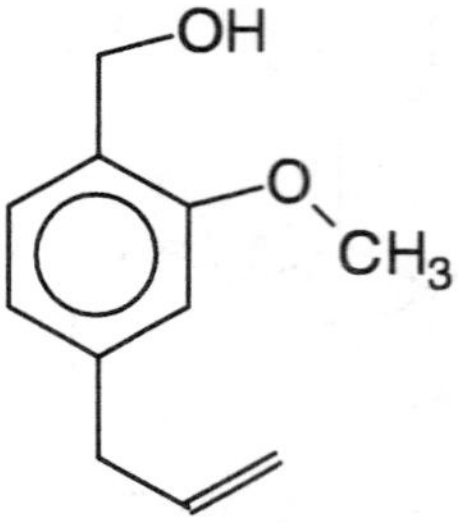

Aldehydes

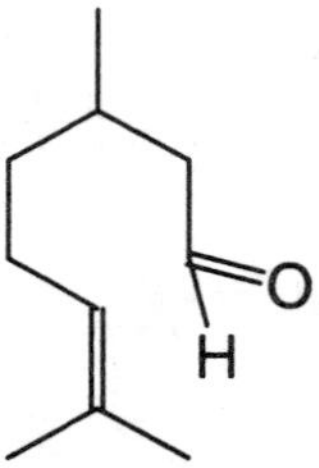

| citronellal | neral | geranial | myrtenal |

| 2-hexenal | decanal | cinnamaldehyde |

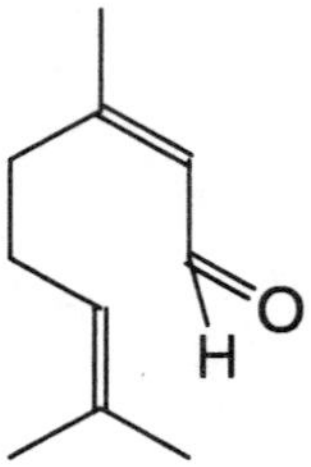

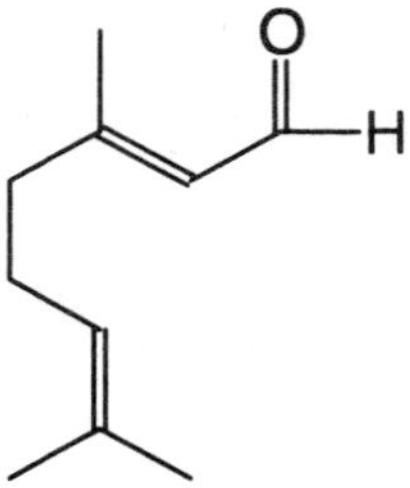

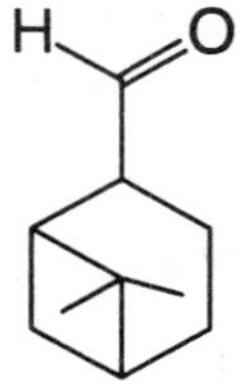

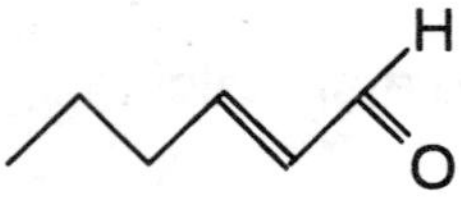

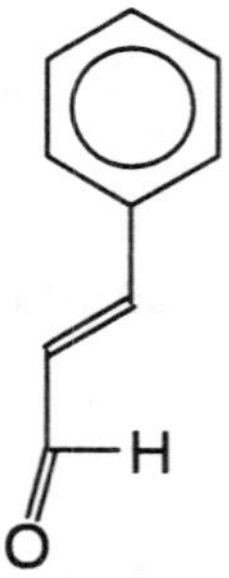

Ketones

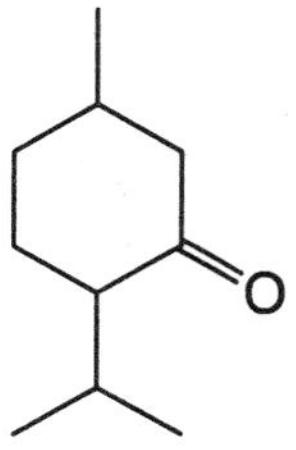

menthone 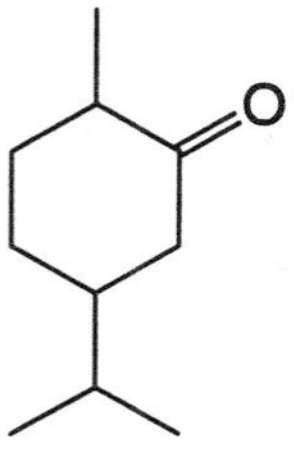   isomenthone 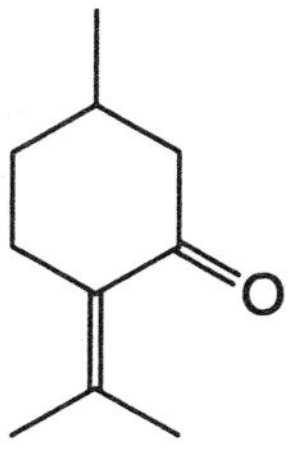   pulegone 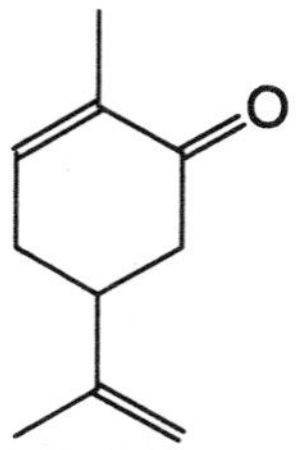   carvone 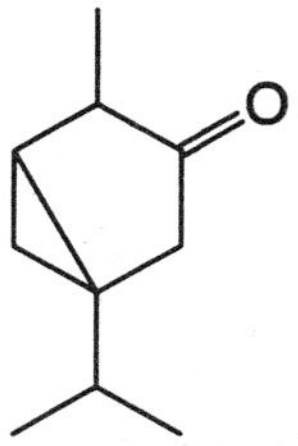   thujone

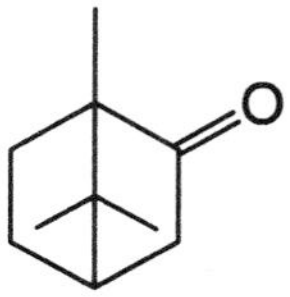

camphor 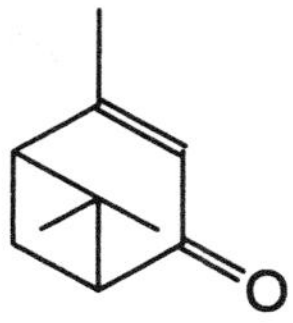   verbenone 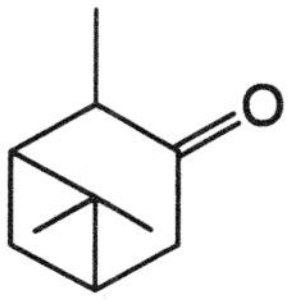   pinocamphone 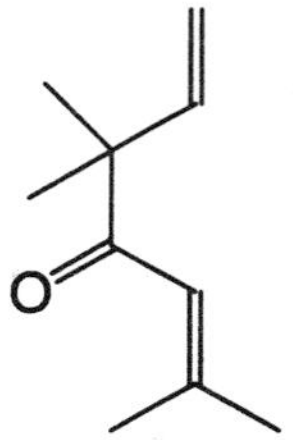   artemisia ketone 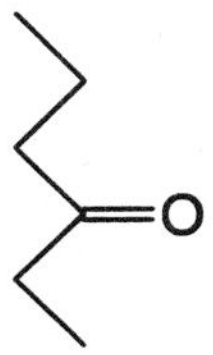   3-hexanone

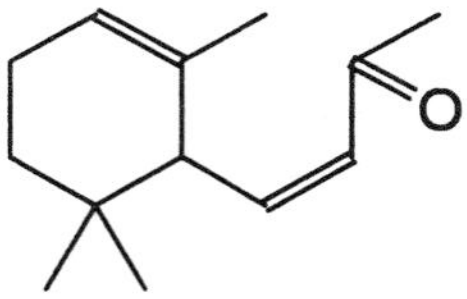

beta-ionone 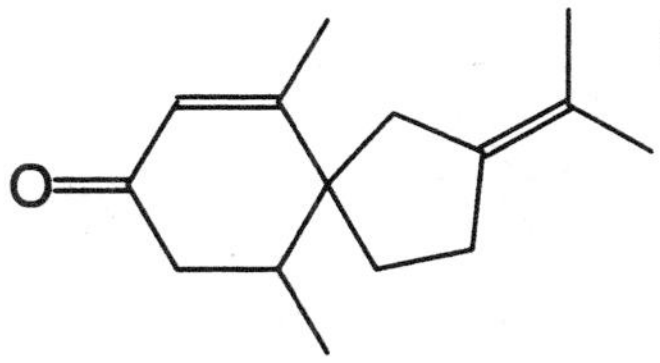   beta-vetivone 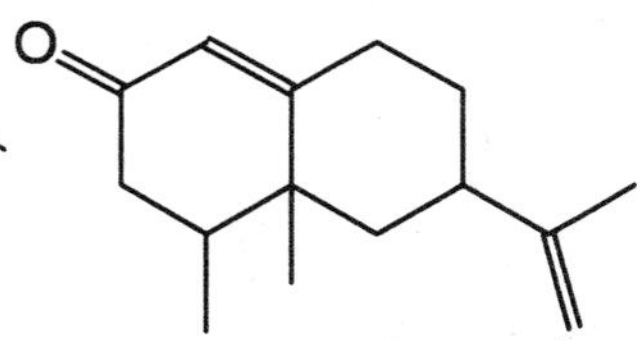   nootkatone

Esters

citronellyl acetate   linalyl acetate   neryl acetate   geranyl acetate

lavandulyl acetate   bornyl acetate   fenchyl acetate   alpha-terpinyl acetate

menthyl butyrate   benzyl benzoate   methyl salicylate

isobutyl angelate   methyl N-methyl anthranilate

Ethers

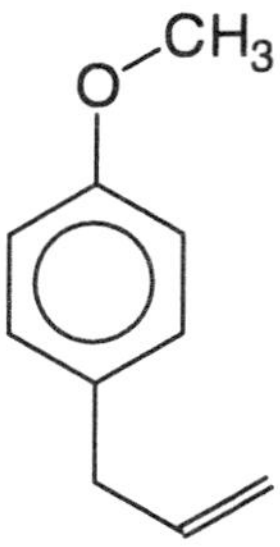

methyl chavicol 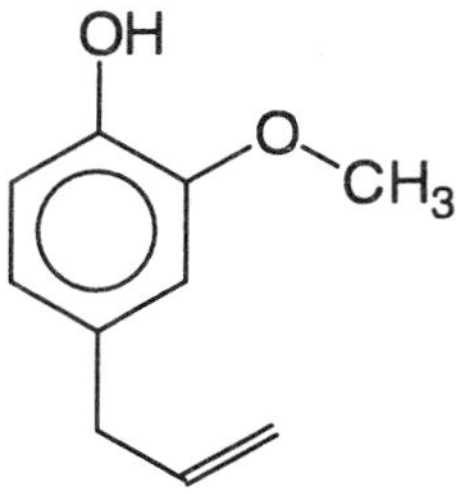   eugenol 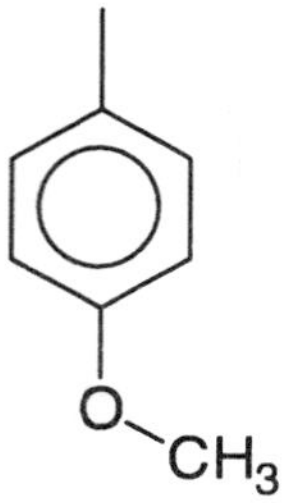   p-cresyl methyl ether 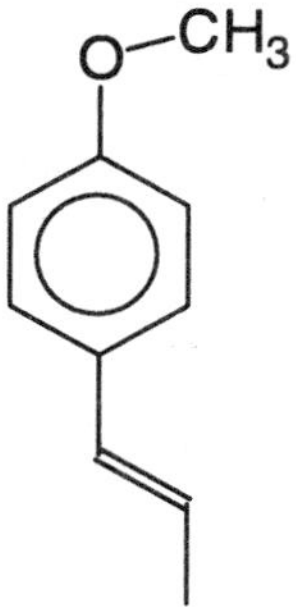   trans-anethole

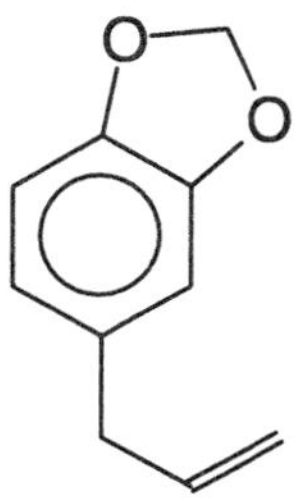

safrole 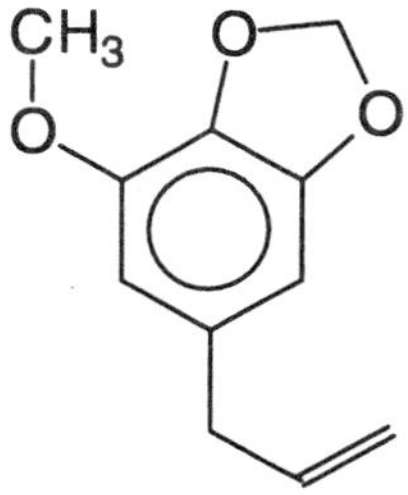   myristicine 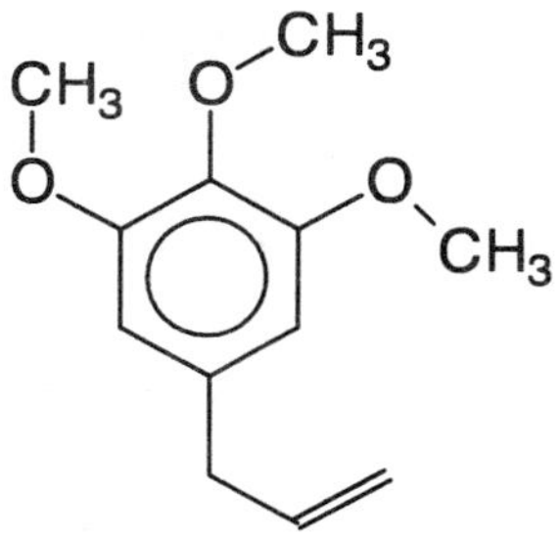   elemicine

Oxides

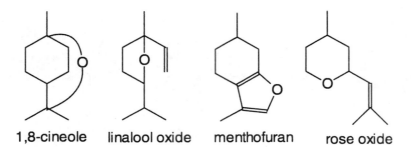

1,8-cineole   linalool oxide   menthofuran   rose oxide

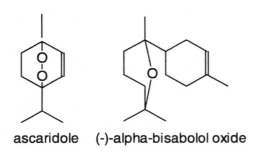

ascaridole   (-)-alpha-bisabolol oxide

Lactones

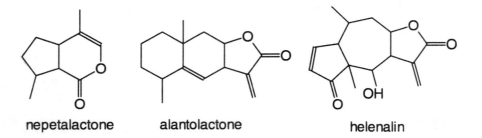

nepetalactone   alantolactone   helenalin

Coumarins

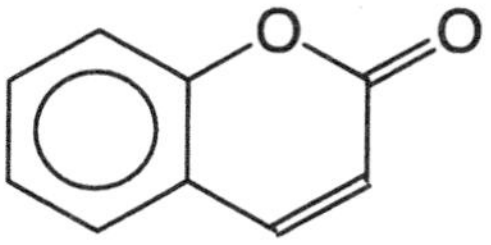

coumarin

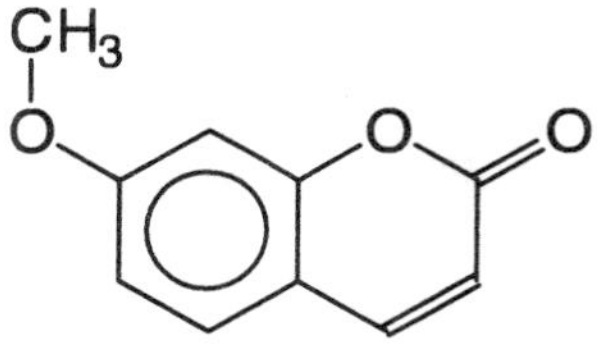

herniarin

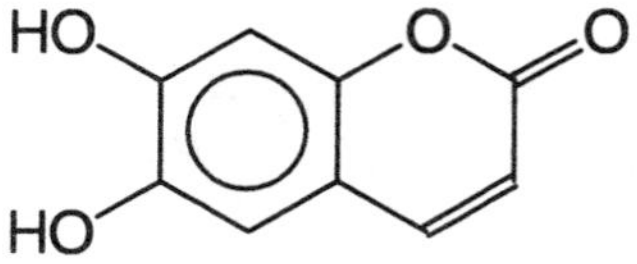

aesculetine

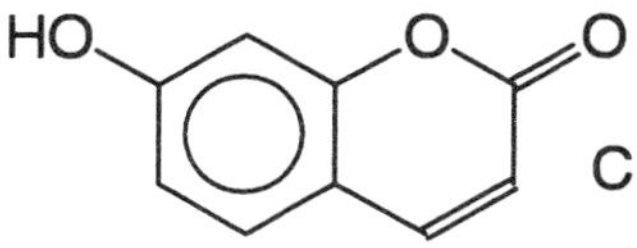

umbelliferone

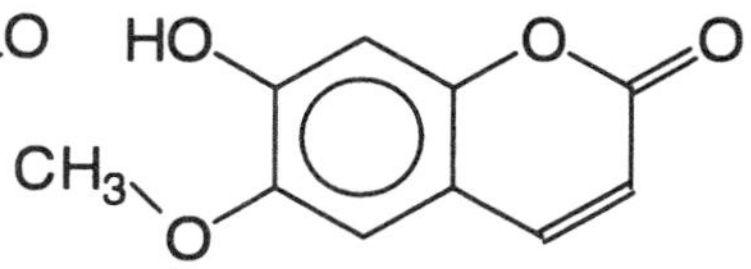

scopoletine

Furocoumarins

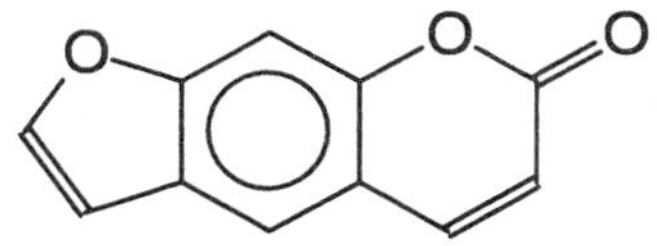

psoralen

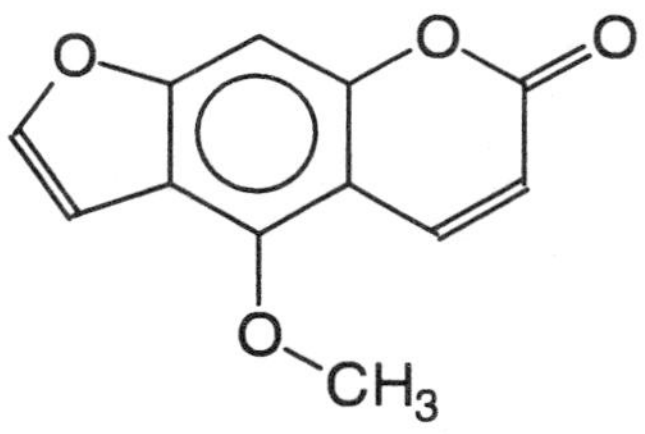

bergapten

RECOMMENDED REFERENCES

Caddy R (1997) *Aromatherapy - Essential Oils in Colour*, Amberwood Publishing, England.

Franchomme P & Pénoël D (1990) *L' aromathérapie exactement,* Roger Jollois, Limoges, France.

Guenther E (1972) *The Essential Oils*, Krieger, Flanders (6 volumes).

Lake M (1989) *Scents and Sensuality*, John Murray Publishers, London.

Sheppard-Hanger S (1997) *The Aromatherapy Practitioner's Reference Manual,* Atlantic Institute of Aromatherapy, USA.

Stoddart D M (1990) *The Scented Ape*, Cambridge University Press, UK.

Tisserand R & Balacs T (1995) *Essential Oil Safety: A guide for health care professionals,* Churchill Livingston, Edinburgh.

Tortora GJ & Grabowski SR (1993) *Principles of Anatomy and Physiology,* HarperCollins College Publishers, USA.

Van Toller S & Todd GH (1988) *Perfumery: The psychology and biology of fragrance*, Chapman & Hall, London.

Journals
Journal of Essential Oil Research, Allured Publishing Corporation, Carol Stream, Illinois, USA.

International Journal of Aromatherapy, Aromatherapy Publications, Sussex, England.

Aromatherapy Today, Aromatherapy Today Publications (Australia), Kellyville, NSW, Australia

Website
Medline database: http://www.ncbi.nlm.nih/PubMed/

INDEX

Essential Oil References from Chapter Six

[1] Chialva F, Monguzzi F, Manitto P, Akgul A (1993)"Essential oil constituents of Trachyspermum copticum fruits" *Journal of Essential Oil Research* Jan/Feb v.5, pp105-106

[2] Srinivas SR (1986) "Composition of Angelica Root oil" in *Atlas of Essential Oils* Ed. SR Srinivas, Bronx, NY

[3] Tabacchi R, Garnero J, Buil R (1974) "Contribution a l'etude de la composition de l'huile essentielle de fruits d'anise de Turque" *Rivista Italiana* v. 56 pp. 683-697.

[4] Chalchat JC, Garry RP, Michet A, Gorunovic M (1991) "Essential oils of Artemisia annua from Yugoslavia" *Rivista Italiana EPPOS* (Special Issue) pp. 471-476

[5] Ronse AC, De Pooter HL (1990) "Essential oil production by Belgian Artemisia alba (Turra) before and after micropropagation" *Journal of Essential Oil Research* Sept/Oct v.2 pp. 237-242

[6] Vernin G et al. (1984) "Analysis of Basil oils by GC-MS Data Bank" *Perfumer & Flavorist* Oct/Nov. v.9 pp. 71-86

[7] Carmo MM, Raposo EJ, Venancio F, Frazao S, Seabra R (1990) "The essential oil of Ocimum basilicum L. from Portugal" *Journal of Essential Oil Research* Sept/Oct v.2 pp. 263-264

[8] McHale D, Laurie WA, Woof MA (1977) "Composition of West Indian Bay Oils" *Food Chemistry* v.2, pp. 19-25

[9] Dugo G, Cotroneo A, Verzera A, Donato MG, del Duce R, Licandro G (1989) "Genuineness characters of the Calabrian Bergamot essential oil" in *Proceedings of 11th International Congress of Essential Oils, Fragrances and Flavours 12-16 Nov, New Delhi, India*, v.4 pp. 245-264

[10] van Gemert LJ, Nijssen LM Maarse H (1983) "Kwaliteitscriteria voor de kruiden Nootmuskaat en Zwarte Peper" TNO-CIVO Food Analysis Institute, private communication (to Boelens)

[11] Klein E, Rojan W (1968) "The most important constituents of Buchu leaf oil" *Dragoco Report* v. 15 pp.3,4

[12] Motl O, Hodacova J, Ubik K (1990) "Composition of Vietnamese Cajeput essential oil" *Flavor & Fragrance Journal* v.5 pp.39-42

[13] Senanayake UN (1977) "The nature, description and biosynthesis of volatiles in Cinnamomum ssp." *Ph.D. thesis,* University of New South Wales, Australia.

[14] Janssen AM (1989) "Antimicrobial activities of essential oils - a pharmacognostical study" *Ph.D. thesis,* Rijksuniversiteit, Leiden, The Netherlands.

[15] Pieribattesi JC, Smadja J, Mondon JM (1988) "Composition of the essential oil of Cardamom from Reunion" in *Flavors and Fragrances: A World Perspective Proceedings of the 10th International Congress of Essential Oils, Fragrance and Flavors, Washington DC, USA 16-20 Nov. 1986* Eds. BM Lawrence, BD Mookherjee, & BJ Willis. Elsevier Science Publishers BV Amsterdam pp. 697-706

[16] De Pooter HL et al. (1988) "The essential oils of five Nepeta species. A preliminary evaluation of the IR use in chemotaxonomy by cluster analysis" *Flavour & Fragrance Journal* 3(4) pp. 155-159

[17] Simon DZ, Beliveau (1987) "Cedarleaf oil (Thuja occidentalis). Extracted by hydro diffusion and steam distillation" *International Journal of Crude Drug Research* v. 25, pp. 4-6

[18] Lawrence BM (1980) "Chemical composition of oils of Cedarwood from Texas and Virginia" *Perfumer & Flavorist* June/July 5(3) p. 63.

[19] Lawrence BM (1980) "Chemical composition of oils of Cedarwood from Texas and Virginia" *Perfumer & Flavorist* June/July 5(3) p. 63.

[20] Wijesekera ROB, Jayewardene AL, Rajapakse LS, Fonseka KH (1974) "Volatile constituents of leaf, stem and root oils of Cinnamon" *Journal of Science, Food and Agriculture* v.25 pp. 1211-1220

[21] Wijesekera ROB, Jayewardene AL, Rajapakse LS, Fonseka KH (1974) op.cit.

[22] Wijesekera ROB (1973) "The chemical composition and analysis of citronella oil" *Journal of the National Science Council (Sri Lanka)* v.1 pp.67-81.

[23] Bruns K, Henirich E, Pagel I (1981) "Citronellaoel: Untersuchung von Handels - und Hybridoelen verschiedener Provenenz" in: *Vorkommen und Analytik aetherischer Ole* Band 2. Eds. KH Cubeczka and G Thieme, Verlag Publishers, Stuttgart. pp?

[24] Lawrence BM (1990) "Comparative chemical composition of commercial Clary Sage oils" *Perfumer & Flavorist* July/Aug v.15 p. 71

[25] Gaydou EM, Randriamiharisoa RP (1987) "Multidimensional analysis of GC data: Application to differentiation of Clove Bud and Clove Stem essential oil from Madagascar" *Perfumer & Flavorist* 12(5), pp.45-51

[26] Kubeczka KH, Schultze W (1987) "Biology and chemistry of conifer oils" *Flavour and Fragrance Journal* 2(4) pp 137-148

[27] Lawrence BM (1980) "The composition of Elemi oil" *Perfumer & Flavorist* Feb/March 5(1) p.57

[28] Boelens MH (1985) "Essential oils and aroma chemicals from Eucalyptus globulus (Labil.)" *Perfumer & Flavorist* 9(6) pp. 1-14

[29] Hunt RS, von Rudloff E (1974) "Chemosystematic studies in the Genus Abies 1. Leaf and twig oil analysis of alpine and balsam firs (average of 10 analyses)" *Canadian Journal of Botany* v.52 pp. 477-487

[30] Vernin G (1989) "GC/MS data bank analysis of the essential oils from Boswellia frereana Birdw. and Boswellia carterii Birdw."in *Flavors and Off-Flavors* Ed. G. Charalambous, Elsevier Science Publishers, Amsterdam pp 511-542.

[31] Abdel Wahab SM et al. (1987) "The essential oil of olibanum" *Planta Medica* v? pp. 382-384

[32] Vernin G, Metzger J, Fraisse D, Sharf C (1983) "Etude des huiles essentielles par CG-SM-Banque SPECMA: Essences de Geranium (Bourbon)" *Parfumerie, Cosmetiques et Aromes* v.52 pp. 51-61

[33] Tsutuslova AL, Antonova RA (1984) "Analysis of Bulgarian daisy oil" *Maslo-Zhir. Prom. St.* v.11 pp 23,24.

[34] Chen CC, Ho CT (1989) "Volatile compounds in ginger oil generated by thermal treatment (steamdistilled oil)" Chapter 34 in: *Thermal generation of aromas*. Eds. TH Parliament, RJ McGorrin, CT Ho, American Chemical Society, Washington DC pp.366-375

[35] Boelens MH (1991) "Critical review on the chemical composition of Citrus oils (normal Israeli grapefruit oil 1988)" *Perfumer and Flavorist* March/April 16 pp.17-34

[36] Lawrence B (1978) "The composition of Ho leaf oil" *Perfumer & Flavorist* 1978.

[37] Joulain D, Ragault M (1976) "Sur quelques nouveaux constituents de l' huile essentielle d'Hyssopus" *Rivista Italiana EPPOS* v. 58 pp. 129-131 and 479-485.

[38] Weyerstahl P et al. (1986) "Isolation and synthesis of compounds from the essential oil of Helichrysum italicum" in *Progress in Essential Oil Research (Proceedings of the International Symposium on Essential Oils)* Ed. EJ Brunke, Walter de Gruyter, Berlin pp.177-195

[39] Srinivas SR (1986) "Chemical composition of Jasmine absolute" in *Atlas of Essential Oils* Ed. SR Srinivas, Bronx, NY pp. 1016-1023.

[40] Kaiser R (1988) "New volatile constituents of Jasminum sambac L. Aiton" in *Flavors and Fragrances: A World Perspective. Proceedings of the 10th International Congress of Essential Oils, Fragrance and Flavors, Washington DC, USA 16-20 Nov. 1986* Eds. BM Lawrence, BD Mookherjee, & BJ Willis. Elsevier Science Publishers BV Amsterdam pp.669-695.

[41] Bonaga G, Galetti GC (1985) "Analysis of volatile components in juniper oil by high resolution gas chromatography and combined gas chromatography/mass spectrometry" *Analytical Chemistry* v. 75 pp. 131-136

[42] Zola A, Le Vanda JP (1979) "Le Lavandin Grosso" *Parfumerie, Cosmetiques et Aromes* Jan/Feb v. 25 pp. 60-62

[43] Le Ster A, Touche J, Linas R, Derbesy M (1986) "Haute-provence French Lavender essential oil" *Proceedings of the 9th International Congress of Essential Oils, Singapore* pp. 127-133

[44] Cappello C et al. (1981) "Richerche chiche sulla composizione dei derivati agrumari Argentini Nota 1. Gli olii essenziali" *Essenze Derivati Agrumari* v.5 pp 229-233

[45] Tittel G, Wagner H, Bos R (1982) "Ueber die chemische Zusammensetzung von Melissenoelen" *Planta Medica* v. 46, pp. 91-98

[46] Haro L, Faas WE (1985) "Comparative study of essential oils of Key and Persian limes (distilled Persian lime)" *Perfumer and Flavorist* Oct/Nov v.10 pp. 67-72

[47] Boelens MH, Jimenez R (1989) "The chemical composition of some Mediterranean Citrus oils" *Journal of Essential Oil Research* v.1 pp 151-159.

[48] Lawrence BM (1981) "The essential oils of Litsea cubeba" *Perfumer & Flavorist* June/July v. 6, p.47

[49] Vostrowsky O, Brosche T, Ihm H, Zintl R, Knobloch K (1981) "Ueber die Komponenten des aetherischen Oelen aus Artemisia absinthum L." *Zeitschrift fur Naturforschung* v.36C pp. 369-377.

[50] Wilson RA, Mookherjee BD (1983) "Characterization of aroma-donating components of Myrrh (headspace analysis)" Paper no. 400, *Proceedings of the 9th International Congress of Essential Oils, Singapore*

[51] Boelens MH, Jimenez R (1991) "The chemical composition of Spanish Myrtle leaf oils Part 1." *Journal of Essential Oil Research* May/June v.3 pp. 173-177

[52] Boelens MH, Jimenez SR (1988) "Essential oils from Seville Bitter Orange (Citrus aurantium L. ssp. amara)" in *Flavors and Fragrances: A World Perspective Proceedings of the 10th International Congress of Essential Oils, Fragrance and Flavors, Washington DC, USA 16-20 Nov. 1986* Eds. BM Lawrence, BD Mookherjee, & BJ Willis. Elsevier Science Publishers BV Amsterdam pp. 551-565.

[53] Ramanolelina PAR et al. (1992) "Chemical composition of Niaouli essential oils from Madagascar (oil of chemotype 1)" *Journal of Essential Oil Research* Nov/Dec v. 4, pp.657-658.

[54] Analytical Methods Committee (1984) "Application of Gas-Liquid Chromatography to the analysis of essential oils Part XI Monographs for seven essential oils" *Analyst* v. 109 pp.1343-1360

[55] Koketsu M, Magaihaes MT, Wilberg VC, Donaliso MGR (1983) "Oleos essenciais de frutos citricos cultivadas no Brazil" *Boletin do. Pesquisas AMBRAPA Centro Technologico Agricolto Alimentario* v.7 p. 21

[56] Ravid U, Putievsky E (1986) "Carvacrol and thymol chemotypes of East Mediterranean wild Labiatae herbs" in: *Progress in Essential Oil Research (Proceedings of the International Symposium on Essential Oils)* Ed. EJ Brunke, Walter de Gruyter, Berlin, pp. 163-167

[57] Randriamihariosa RP, Gaydou EM (1987) "Composition of Palmarosa oil" *Journal of Agriculture and Food Chemistry* v. 35 pp. 62-66.

[58] Srinivas SR (1986) "Composition of Patchouli oil" in *Atlas of Essential Oils* Ed. SR Srinivas, Bronx, NY pp?

[59] Hefendehl FW (1970) "Betirage zur Biogenese aetherische Oele Zusammensetzung zweier Aetherischer Oele von Mentha pulegium L. (C.R. variety)" *Phytochemistry* v.9, pp. 1985-1995

[60] Embong MB, Steele L, Hadziyev D, Molnar S (1977) "Essential oils from herbs and spices grown in Alberta" *Journal d' Institute de Canadien Science de Technologie Alimentaire* 10(4) pp.247-256.

[61] Boelens MH, Jimenez SR (1986) "Essential oils from Seville Bitter Orange" *Flavors and Fragrances: A World Perspective. Proceedings of the 10th International Congress of Essential Oils, Fragrance and Flavors, Washington DC, USA 16-20 Nov. 1986* Eds. BM Lawrence, BD Mookherjee, & BJ Willis. Elsevier Science Publishers BV Amsterdam pp. 551-565

[62] Kubeczka KH, Schultze W (1987) op. cit.

[63] Kubeczka KH, Schultze W (1987) op. cit.

[64] Hasebe A, Oomura T, (1989) "The constituents of essential oils from Anthemis nobilis" *Koryo* v 161 pp. 93-101

[65] Garnero J, Guichard G, Buil P (1976) "L'huile essentielle et la concentre de rose de Turquie" *Parfumerie, Cosmetiques et Savons* v. 8 pp. 33-46

[66] Karawya MS, Hashim FM, Hifnawy MS (1974) "Oils of Jasmin, Rose and Cassie of Egyptian Origin" *Bulletin of the Faculty of Pharmacy, University of Cairo* v 13 pp. 183-192

[67] Boelens MH (1985) "The essential oils from Rosmarinus officinalis L." *Perfumer & Flavorist* Oct/Nov v.10 pp 21-37

[68] Fournier G, Habib J, Reguigui A, Safta F, Guetari S, Chemli R (1989) "Etude de divers echantillons d'huile essentielle de Romarin de Tunisie" *Plantes Medicinales et Phytotherapie* v. 23 pp.180-185

[69] Formacek K, Kubeczka KH (1982) "The chemical composition of a commercial Bois-de-Rose oil" in *Essential oils analysis by capillary chromatography and carbon-13 NMR spectroscopy* J. Wiley & Sons, NY,

[70] Aboutab EA, Elazzouny AA, Hammerschmidt FJ (1988) "The essential oil of Ruta graveolens L. growing in Egypt" *Pharmacological Science* v.56, pp. 121-124

[71] Vernin G, Metzger J (1986) "Analysis of Sage oils by GC-MS Data bank" *Perfumer & Flavorist* 11(5) pp. 79-84

[72] Brunke EJ, Hammerschmidt FJ (1988) "Constituents of East Indian Sandalwood oil. An Eighty year long "stability test" (concentration in fresh oil)" *Dragoco Report* n.4 pp 107-133

[73] Chialva F, Liddle PAP, Ulian F, de Smedt P (1980) "Indagine sulla composizione dell olio ess. di Satureja hortensis L. coltivata in Piemonte e confronto con altr di diversa origine" *Rivista Italiana* v.62 pp.297-300

[74] Kokkikini S, Vokou D (1989) "Mentha spicata (Lamiaceae) chemotypes grown wild in Greece" *Economic Botany* 43(2) pp. 192-202

[75] de Teresa P et al. (1989) "Chemical composition of the Spanish Spike oil (lab-distilled sample)" *Planta Medica* v. 55 p.398

[76] Cu JQ, Perineau F, Goepfort G (1990) "GC/MS analysis of Star Anise oil" *Journal of Essential Oil Research* Mar/Apr v.2 pp. 91-92

[77] Akgul A (1986) "Studies on the essential oils from Turkish fennel seeds" in: *Progress in Essential Oils (Proceedings of the International Symposium on Essential Oils)* Ed. EJ Brunke, Walter de Gruyter, Berlin pp. 487-489

[78] Oberdieck R (1981) "Ein Beitrag zur Kenntnis und Analytik von Majoran (Marjorana hortensis Moench)" *Zeitschrift fur Naturforschung* v.36 pp. 23-29

[79] De Pooter HL, Vermeesch J, Schamp NM (1989) "The essential oils of Tanacetum vulgare L. and Tanacetum parthenium L. SchultzpBip" *Journal of Essential Oil Research* 1(1) pp. 9-13

[80] Tucker AO, Maciarello MJ (1987) "Plant identification" in: *Proceedings of the First National Herb Growing and Marketing Conference* Eds. JE Simon, L Grant, Purdue University Press, West Lafayette IN, pp. 126-172.

[81] Williams LR, Home VN (1989) "Plantations of Melaleuca alternifolia - A revitalised Australian Tea Tree oil industry" in *Proceedings of the 11th International Congress of Essential Oils, Fragrances and Flavours, 12-16 Nov.* v.3 pp.49-53

[82] Piccaglia R, Marotti M (1991) "Composition of the essential oil of an Italian Thymus vulgaris L. ecotype" *Flavour and Fragrance Journal* v.6, pp.241-244

[83] Zwaving JH, Bos R (1992) "Analysis of the essential oils of five Curcuma species" *Flavor and Fragrance Journal* v.7, pp. 19-22

[84] Klimes I and Lamparsky D (1976) "Vanilla volatiles. A comprehensive analysis" *International Flavour and Food Additives* v.7 pp. 292-273

[85] Garnero J (1972) "A survey of the vetiver oil components (Percentages estimated)" *Rivista Italiana EPPOS* v.54 p. 315

[86] Frey C (1988) "Detection of synthetic flavorant addition to some essential oils by selected ion monitoring GC/MS (estimated peak percentages)" in *Flavors and Fragrances: A World Perspective. Proceedings of the 10th International Congress of Essential Oils, Fragrance and Flavors, Washington DC, USA 16-20 Nov. 1986* Eds. BM Lawrence, BD Mookherjee, & BJ Willis. Elsevier Science Publishers BV Amsterdam

[87] Falk AJ, Bauer L, Bell CL (1974) "The constituents of the essential oil of Achillea millefolium" *Llodia* v. 37 pp. 598-602.

[88] Gaydou EM, Randriamiharisoa R, Bianchini JP (1986) "Composition of the essential oil of Ylang-ylang (Cananga odorata Hook F. et Thomson)" *Journal of Agriculture and Food Chemistry* v.34 pp. 481-487.